SpringerBriefs in Petroleum Geoscience & Engineering

Series editors

Dorrik Stow, Heriot-Watt University, Edinburgh, UK

Mark Bentley, AGR TRACS International Ltd, Aberdeen, UK

Jebraeel Gholinezhad, University of Portsmouth, Portsmouth, Hampshire, UK

Lateef Akanji, King's College, University of Aberdeen, Scotland, UK

Khalik Mohamad Sabil, Heriot-Watt University, Putrajaya, Malaysia

Susan Agar, Houston, USA

Kenichi Soga, Department of Civil and Environmental Engineering, University of California, Berkeley, CA, USA

A. A. Sulaimon, Department of Petroleum Engineering, Universiti Teknologi Petronas, Seri Iskandar, Perak, Malaysia

The SpringerBriefs series in Petroleum Geoscience & Engineering promotes and expedites the dissemination of substantive new research results, state-of-the-art subject reviews and tutorial overviews in the field of petroleum exploration, petroleum engineering and production technology. The subject focus is on upstream exploration and production, subsurface geoscience and engineering. These concise summaries (50–125 pages) will include cutting-edge research, analytical methods, advanced modelling techniques and practical applications. Coverage will extend to all theoretical and applied aspects of the field, including traditional drilling, shale-gas fracking, deepwater sedimentology, seismic exploration, pore-flow modelling and petroleum economics. Topics include but are not limited to:

- Petroleum Geology & Geophysics
- Exploration: Conventional and Unconventional
- Seismic Interpretation
- Formation Evaluation (well logging)
- Drilling and Completion
- Hydraulic Fracturing
- Geomechanics
- Reservoir Simulation and Modelling
- Flow in Porous Media: from nano- to field-scale
- Reservoir Engineering
- Production Engineering
- Well Engineering; Design, Decommissioning and Abandonment
- Petroleum Systems; Instrumentation and Control
- Flow Assurance, Mineral Scale & Hydrates
- Reservoir and Well Intervention
- Reservoir Stimulation
- Oilfield Chemistry
- Risk and Uncertainty
- Petroleum Economics and Energy Policy

Contributions to the series can be made by submitting a proposal to the responsible Springer contact, Charlotte Cross at charlotte.cross@springer.com or the Academic Series Editor, Prof Dorrik Stow at dorrik.stow@pet.hw.ac.uk.

More information about this series at http://www.springer.com/series/15391

Dayanand Saini

CO_2-Reservoir Oil Miscibility

Experimental and Non-experimental Characterization and Determination Approaches

 Springer

Dayanand Saini
Department of Physics and Engineering
California State University
Bakersfield, CA, USA

ISSN 2509-3126   ISSN 2509-3134   (electronic)
SpringerBriefs in Petroleum Geoscience & Engineering
ISBN 978-3-319-95545-2   ISBN 978-3-319-95546-9   (eBook)
https://doi.org/10.1007/978-3-319-95546-9

Library of Congress Control Number: 2018948131

Printed on acid-free paper

This Springer imprint is published by the registered company Springer International Publishing AG part of Springer Nature
The registered company address is: Gewerbestrasse 11, 6330 Cham, Switzerland

I dedicate this work to my good friend and former fellow student Daryl S. Sequeira, who guided and mentored me through my first exposure to the topic of CO_2-reservoir oil miscibility and its experimental characterization and determination in terms of minimum miscibility pressure (MMP) during my early days as a graduate student at the Craft and Hawkins Department of Petroleum Engineering, Louisiana State University (LSU), Baton Rouge. This work is also dedicated to my Ph.D. Advisor, Dr. Dandina Rao, Emmett C. Wells Jr. Distinguished Professor of Petroleum Engineering at LSU, who provided me with many opportunities to expand my knowledge of the phenomenon of CO_2-reservoir oil miscibility.

Preface

A reliable knowledge of the minimum miscibility pressure (MMP) of carbon dioxide (CO_2) plays a key role in developing cost-effective and efficient miscible CO_2 flooding projects for both enhanced oil recovery (EOR) and simultaneous EOR and storage purposes. Hence, interested students and industry professionals who are either planning to join or are already involved in such projects will greatly benefit from a text that can enhance their understanding of both the phenomenon of CO_2-reservoir oil miscibility and the various experimental and non-experimental approaches that are used to characterize and determine it in terms of MMP. This book is meant to serve that very purpose.

Bakersfield, USA   Dayanand Saini

Acknowledgements

I would like to thank the Department of Physics and Engineering and the School of Natural Sciences, Mathematics, and Engineering (NSME) for providing me an opportunity to continue my research in the area of CO_2-reservoir oil miscibility at the California State University, Bakersfield (CSUB).

The research performed at CSUB during 2013–16 had later prompted me to publish two publications: "Modeling of Pressure Dependence of Interfacial Tension Behaviors of a Complex Supercritical CO_2 + Live Crude Oil System Using a Basic Parachor Expression" in *the Journal of Petroleum and Environmental Biotechnology* (Vol 7: 277) and "An Investigation of the Robustness of Physical and Numerical Vanishing Interfacial Tension Experimentation in Determining CO_2 + Crude Oil Minimum Miscibility Pressure" in *the Journal of Petroleum Engineering* (Vol 2016, Article ID 8150752). The contribution of the above-mentioned publications in developing the content for this book is duly acknowledged.

A special thanks to Ms. Charlotte Cross, Assistant Editor (Engineering), Springer London, for her time, effort, and encouragement that made the publication of this book a reality. Last but not least, thanks to my daughter, Lalima, and my wife, Rekha, for their unwavering support and understanding during the preparation of the manuscript.

Contents

Abbreviations

1D	One-Dimensional
AMS	Acoustically Monitored Separator
ARI	Advanced Resources International
CCUS	Carbon Capture, Use, and Storage
CO_2	Carbon Dioxide
DIM	Diminishing Interface Method
DOE	Department of Energy
EOR	Enhanced Oil Recovery
EOS	Equation of State
FC-MMP	First Contact Minimum Miscibility Pressure
GA	Genetic Algorithm
H_2S	Hydrogen Sulfide
HCPV	Hydrocarbon Pore Volume
HPHT	High-Pressure High-Temperature
ICA	Imperialistic Competitive Algorithm
IFT	Interfacial Tension
LGT	Linear Gradient Theory
LPG	Liquefied Petroleum Gas
LSIPs	Large-Scale Integrated Carbon Capture and Storage Projects
MCM	Multicontact Miscibility
MC-MMP	Multicontact Minimum Miscibility Pressure
MCT	Multicontact
MMP	Minimum Miscibility Pressure
MMV	Monitoring, Management, and Verification
MOC	Method of Characteristics
MVA	Monitoring, Verification, and Accounting
NETL	National Energy Technology Laboratory
OXY	Occidental Petroleum Corporation
P/T	Pressure/Temperature
PC-SAFT	Perturbed-Chain Statistical Associating Fluid Theory

PREOS	Peng–Robinson Equation of State
PVT	Pressure, Volume, Temperature
RBA	Rising Bubble Apparatus
RBF	Radial Basis Function
ROZ	Residual Oil Zone
RPI	Rapid Pressure Increase
SACROC	Scurry Area Canyon Reef Operators Committee
$ScCO_2$	Supercritical Carbon Dioxide
SFE	Supercritical Fluid Extraction
SRM	Sonic Response Method
UCSP	Upper Critical Solution Pressure
VIT	Vanishing Interfacial Tension
VLE	Vapor–Liquid Equilibrium
WAG	Water Alternating Gas

Chapter 1
Fundamentals of CO_2-Reservoir Oil Miscibility

Abstract The efficient and coeffective implementation and operation of CO_2 injection-based enhanced oil recovery (CO_2-EOR) and/or simultaneous CO_2-EOR and storage projects often rely on accurate characterization and determination of the phenomenon of CO_2-reservoir miscibility in terms of the minimum miscibility pressure (MMP). Hence, for a better understanding of both the CO_2-reservoir oil miscibility and the MMP it is necessary that various fundamental aspects of CO_2-reservoir oil miscibility are understood first. Different approaches, including molecular dynamics-based, interfacial energy-based, and thermodynamics-based approaches can be used to better understand the complex interactions occurring between injected CO_2 and reservoir oil that ultimately lead to the development of miscibility at the MMP. To enhance our understanding of the phenomenon of CO_2-reservoir oil miscibility, a discussion of the key differences between solubility and miscibility and the solvent properties of supercritical CO_2 is also presented.

1.1   Introduction

The beneficial role of high miscibility between carbon dioxide (CO_2) and reservoir oil at pressures more than 6.8 MPa (1,000 psi) was clearly recognized by Whorton et al. (1952) when they filed the first U.S. patent for a method for producing oil by means of CO_2 in the 1950s. Later, a large-scale commercial enhanced oil recovery (EOR) project, which involved the injection of CO_2 for recovering more oil from depleted reservoirs, was first implemented in the Permian Basin of West Texas by the Scurry Area Canyon Reef Operators Committee (SACROC) Unit of the Kelly-Snyder Field and North Cross floods in the early 1970s (Meyer 2007; Global CCS Institute 2011). These innovations opened the door for CO_2 injection-based miscible flooding techniques to become among the most popular EOR methods to date for improving overall oil recovery factors in conventional reservoirs.

According to Qin et al. (2015), there are 152 CO_2-EOR projects around the world (as per 2014 data) and majority of them (137) are in the United States. Out of 137 US CO_2-EOR projects, 128 projects are miscible flooding projects accounts. As

D. Saini, *CO₂-Reservoir Oil Miscibility*, SpringerBriefs in Petroleum Geoscience & Engineering, https://doi.org/10.1007/978-3-319-95546-9_1

both the data and the name suggest, CO$_2$-reservoir oil miscibility plays a key role in the success of CO$_2$ injection-based miscible flooding EOR as well as simultaneous CO$_2$-EOR and storage projects in conventional reservoirs. However, a knowledge of the fundamentals (theory, scientific principles, and engineering applications) of CO$_2$-reservoir oil miscibility is desired for a better understanding of its role in the successful design, implementation, and operation of a miscible CO$_2$-EOR project. It is worth mentioning here that in contrast to unconventional reservoirs where certain reservoir and reservoir fluids characteristics (i.e., low permeability, porosity, and oil viscosity) require the use of specialized drilling (e.g., horizontal wells), completion (e.g., hydraulic fracturing), and production (e.g., steamflooding) techniques for commercial exploitation of hydrocarbon reserves, conventional reservoirs use relatively simple and standard drilling, completion, and production methods for producing hydrocarbons in commercial quantities.

Before the fundamentals of CO$_2$-reservoir oil miscibility are discussed in detail, we need to appreciate the fact that it is an injection fluid of choice for the petroleum industry. Obviously, it shows high miscibility with reservoir oils; however, it is also less expensive than other similarly miscible fluids (NETL 2010) such as nitrogen and liquified petroleum gas (LPG). On another hand, a significant portion of injected CO$_2$ is lost in the reservoir anyway. This incidental storage of injected CO$_2$ in an EOR project, in the form of carbon capture, utilization, and sequestration (CCUS), provides another value-added benefit, and the industry has already seized upon it. The recent significant growth in the portfolio of large-scale integrated carbon capture and storage projects (LSIPs), which mainly rely on the utilization of commercially-captured anthropogenic CO$_2$ for maximizing both aspects (i.e., oil recovery and CO$_2$ storage) of miscible CO$_2$-EOR projects, clearly indicates the increased importance and acceptance of this value-added benefit. A detailed discussion of the synergy of EOR and storage in miscible CO$_2$-EOR operations can be found elsewhere (Saini 2017).

1.2   Miscibility

Before discussing the specifics of the tools and techniques that are current available for characterizing and determining CO$_2$-reservoir oil miscibility, a brief discussion on the basic definition and general theory of miscibility is necessary. The term *miscibility* is often used to refer the ability of a liquid solute to dissolve in a liquid solvent, whereas, solubility is a more general term, often referring to the ability of a solid solute to dissolve in a liquid solvent (Suico 2017). Hence, miscibility, a qualitative rather than a quantitative observation, means how completely two or more liquids dissolve in each other (Net Industries 2017). Formally, miscibility or infinite mutual solubility can be described as *"physical condition between two fluids that will permit them to mix in all proportions without an interface being formed by the materials"* (Clark et al. 1958).

Miscible liquids generally mix without limit, meaning they are soluble at all amounts (Suico 2017). For example, two common liquids—namely water and ethyl alcohol—are completely miscible whether the solution is 1% water and 99% ethyl alcohol, 50% of both, or 1% ethyl alcohol and 99% water (Net Industries 2017). In this example, mixing (i.e., mutual solubility) of water and ethyl alcohol is independent of their individual concentrations. Hence, they have infinite mutual solubility and are completely miscible. However, it was recognized long ago, that without a knowledge of the law of molecular attraction, the required conditions for complete miscibility or infinite solubility cannot be dealt with quantitatively (Tyrer 1912).

The extent of miscibility between two liquids can be explained on the basis of molecular attractions. According to Burke (1984), a certain amount of intermolecular stickiness holds the molecules of the liquid together. This intermolecular stickiness or sticky forces between molecules are called van der Waals forces, which are the result of electromagnetic interactions between the molecules (Burke 1984). Furthermore, compared to a neutral (i.e., neither positively charged nor negatively charged) molecule, an individual molecule itself may possess a different electron shell density, which turns a molecule in a small magnet, or dipole. The extent of the deviations in electron shell density depends on the physical architecture of the molecule, hence, certain molecular geometries may result in strongly polar substances, while other configurations may result in only a weakly polar substance (Burke 1984). Thus, the substances that have similar polarities are easily soluble in each other, but as, the deviations in polarity increases, solubility becomes increasingly difficult. As further discussed by Burke (1984), the extent of solubility not only depends on the different degrees of similar polarities, but it is also affected by the degree of similarity between other intramolecular interactions such as hydrogen bonding, induction and orientation effects, and dispersion forces.

Normally, polar liquids (solute molecules) tend to mix easily with polar liquids (solvent molecules), because the dipole-dipole attractions (or hydrogen bonding) of the solute molecules are at least as strong as those between molecules in the pure solute or in the pure solvent (OpenStax 2016). Similarly, non-polar liquids mix easily with each other because there is no appreciable difference in the strengths of solute-solute, solvent-solvent, and solute-solvent intermolecular attractions (OpenStax 2016). Thus, polar molecules in polar solvents and nonpolar molecules in nonpolar solvents show miscible behavior. An example of miscible behavior between two fluids (e.g., water and isopropyl alcohol) is demonstrated in Fig. 1.1. As can be seen in Fig. 1.1, when colored water (Fig. 1.1a) and dyed isopropyl alcohol (Fig. 1.1b) are mixed, they immediately form a miscible phase (i.e., a dark colored solution shown in Fig. 1.1c).

Certain liquids, such as bromine and water, which show moderate mutual solubility are said to be partially miscible (OpenStax 2016). In case of bromine and water, when mixed together, two distinct layers, a top layer containing saturated solution of bromine in water and a bottom layer consisting a saturated solution of water in bromine, can be easily distinguished. Bromine, being a nonpolar liquid shows limited solubility (i.e., partial miscibility) in water, which is the most common polar solvent. On the other hand, many non-polar liquids (e.g., hydrocarbons such as benzene,

(a)   Colored water   (b)   Dyed isopropyl alcohol   (c)   Dark miscible phase formed by
                                                                                   colored water and dyed isopropyl
                                                                                   alcohol (dark) phases

Fig. 1.1 Example of miscible (complete mutual solubility) behavior between colored water and
dyed isopropyl alcohol (dark) phases

dodecane etc.) do not mix to an appreciable extent with polar liquids such as water.
Because they show very low (i.e., non-zero) mutual solubility, they are called immiscible. The attraction between the molecules of such non-polar liquids and polar water
molecules is ineffectively weak and the only strong attractions in such a mixture are
between the water molecules, so they effectively squeeze out the molecules of the
non-polar liquid (OpenStax 2016). The immiscible behavior, as shown by mineral
oil (Fig 1.2a) and isopropyl alcohol (Fig 1.2b) phases by forming two distinct phases
(Fig. 1.2c), is clearly visible in Fig. 1.2.

As further summarized by OpenStax (2016), the distinction between immiscible
and miscible fluids is essentially the extent of miscibility as miscible liquids are of
infinite mutual solubility, while liquids said to be immiscible are of very low (though
not zero) mutual solubility. Also, in case of miscibility, conventional concepts of
solute (the minor component in a solution) and solvent (the liquid in which solute is
dissolved to form a solution) do not matter as miscibility itself is a manifestation of
infinite solubility, which is independent of the relative concentrations of individual
liquids forming the solution. The same is true for a solution resulting from the mixing
of multiple liquids.

In nutshell, solubility is determined by the balance of intermolecular forces
between solute and solvent as well as the entropy change that occurs on the dis-
persal of the solute in the solvent, thereby depending on the system's pressure and
temperature (Peach and Eastoe 2014). Obviously, a system's total volume remains
constant when two liquids are mixed to form a solution. On the other hand, when
it comes to mutual solubility (i.e., miscibility) of CO$_2$ and reservoir oils, first, we
need to appreciate the fact that the injection of CO$_2$ for EOR purposes is carried
out at the conditions of elevated pressure (i.e., injection pressure) and temperature

(a) Mineral oil   (b) Dyed isopropyl alcohol   (c) Immiscible mineral oil (clear)
                                                                   and dyed isopropyl alcohol
                                                                   (dark) phases

Fig. 1.2 Example of immiscible (a very low mutual solubility) behavior between mineral oil (clear)
and dyed isopropyl alcohol (dark) phases

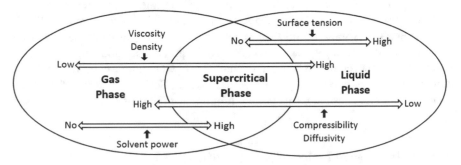

Fig. 1.3 Variation in key characteristics of CO_2 in different states (i.e., gas phase, supercritical
phase, and liquid phase)

(i.e., reservoir temperature). These pressure and temperature conditions are always
well beyond the critical pressure and critical temperature of CO_2. When placed at
a pressure and temperature above its critical pressure and critical temperature (i.e.,
in a supercritical state), CO_2 exhibits certain unique, interesting, and useful characteristics (Fig. 1.3) including gas-like diffusivity and liquid-like solvent densities.
Secondly, CO_2 is generally considered as a non-polar solvent, primarily because of its
zero-molecular dipole moment and its low dielectric constant; however, it may also
show polar attributes (Raveendran et al. 2005). In contrast, reservoir oils, because
they are a complex mixture of substances, may show a variety of polarities ranging
from extremely nonpolar, low polar, polar, highly polar, extrapolar (Bruening 1991).
     Nevertheless, supercritical CO_2 (ScCO_2) is one of the most commonly used sol-
vent as well as displacing agent (like other oil solvents and displacing agents such as

nitrogen and LPG) in miscible flooding projects meant to improve oil recovery. However, the condition of complete miscibility (i.e., infinite solubility) between CO$_2$ (i.e., ScCO$_2$) and a given reservoir oil occurs at certain operating pressure and reservoir temperature. Depending on the nature of contacts (i.e., first contact or multiple contacts) occurring between injected CO$_2$ and reservoir oil for establishing miscibility, the pressure is referred to as first contact-minimum miscibility pressure (FC-MMP) or multiple contact-minimum miscibility pressure (MC-MMP). The MC-MMP is simply referred as the MMP throughout the text.

The first-contact miscibility refers to a condition wherein injected CO$_2$ and reservoir oil become miscible at their first ever contact. The phenomenon of achieving miscibility through first-contact process is generally exhibited by solvents like LPG and light oils; however, CO$_2$ cannot achieve first-contact miscibility in most oil reservoirs within a reasonable range of pressures and reservoir temperature. In the case of a given reservoir oil and injected CO$_2$ system, for complete miscibility to occur, multiple contacts between the two phases are needed. In a multicontact process, CO$_2$ and various components of reservoir oil make numerous contacts by moving back and forth within the contacting phases (i.e., the CO$_2$-rich phase and the oil-rich phase) until a completely miscible phase (i.e., a homogeneous phase) is formed. The lowest operating pressure at which condition of multicontact miscibility between injected CO$_2$ and reservoir oil occur is termed as minimum miscibility pressure (MMP).

Whether a CO$_2$-reservoir oil system achieves miscibility through the process of first-contact or miscibility is achieved through the process of multicontact, it is necessary that, for developing a better understanding of CO$_2$-reservoir oil miscibility at fundamental level, one look at the phase behavior of ScCO$_2$ and some of the basic characteristics responsible for its infinite ranging and inherent solvent properties.

1.3 Solvent Properties of Supercritical CO$_2$

As can be seen in Fig. 1.4, a supercritical region is the region of CO$_2$ phase diagram, which is beyond the critical point of CO$_2$ (i.e., critical pressure of 7.38 MPa (1,070.3 psi) and critical temperature of 30.98 °C (87.76 °F)). In a supercritical state, which is easily achieved in such processes a CO$_2$-injection-based EOR operations, CO$_2$ possesses both liquid-like (high density) and gas-like (low viscosity and high diffusivity) properties as well as an intermediate or transitional behavior between that of a liquid and a gas (Supercriticalfluid website 2017), recognized by the widom region (Fig. 1.4). Certain properties of ScCO$_2$ can be exploited for many useful scientific purposes and industrial processes (Kaziunas and Schlake 2016). In case of conventional liquid solvents such as propane and other supercritical fluids such as nitrogen, due to their relatively low compressibilities, very large pressures are required to make any appreciable change in their densities, whereas, a significant change in the density of ScCO$_2$ can be obtained by a change in pressure (Fig. 1.5).

It is also well known that all the thermodynamic properties of a pure substance, in our case ScCO$_2$, can be obtained by the derivative form of a fundamental equation

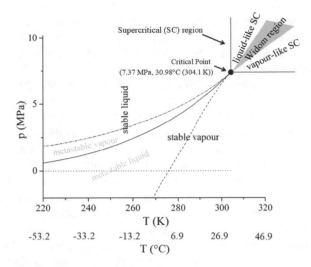

Fig. 1.4 CO$_2$ phase diagram, reprinted with modification by permission from Copyright Clearance Center: Springer Nature, Environmental Earth Sciences, Anomalous fluid properties of carbon dioxide in the supercritical region: application to geological CO$_2$ storage and related hazards, Imre AR, Ramboz C, Deiters UK et al., © 2015

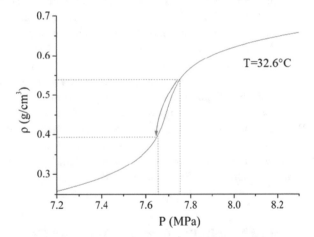

Fig. 1.5 Example of significant change (more than 25%) in supercritical CO$_2$ density due to an isothermal pressure drop (T = 32.6 °C) of 0.1 MPa, reprinted with modification by permission from Copyright Clearance Center: Springer Nature, Environmental Earth Sciences, Anomalous fluid properties of carbon dioxide in the supercritical region: application to geological CO$_2$ storage and related hazards, Imre AR, Ramboz C, Deiters UK et al., © 2015

of state explicit in the Helmholtz energy, which is a function of two independent variables—density and temperature (Span and Wagner 1996). On the other hand, we know that, like its other thermodynamic properties, the solvating power of ScCO$_2$ can also be easily adjusted by manipulating the pressure and temperature conditions of the system. However, there are certain basic characteristics that need to be considered for a fundamental understanding of the solvating power of ScCO$_2$. Raveendran et al. (2005) have provided a detailed discussion of the fundamental nature of CO$_2$ (i.e., ScCO$_2$) as a solvent, which is summarized below.

According to Raveendran et al. (2005), a more molecular level "chemical" description of the solvation behavior of CO$_2$ suggests that CO$_2$ can act as a weak Lewis acid (i.e., a substance, such as H$^+$ ion, with capability to accept a pair of nonbonding electrons) as well as a Lewis base (i.e., a substance, such as OH$^-$ ion, that can donate a pair of nonbonding electrons). Also, various theoretical and experimental evidence indicates that CO$_2$ can participate in hydrogen-bonding interactions (Raveendran et al. 2005). Hence, at a fundamental level, CO$_2$ needs to be considered as a polar solvent, and its solvent properties depend on the solute-solute, solute-solvent, and solvent-solvent interactions. It is the nature of site-specific solute-solvent interactions (i.e., how easily the pairwise solute-solute interaction energies can be overcome by the pairwise solute-solvent interaction energies), that ultimately decides the solubility of materials in CO$_2$ (Raveendran et al. 2005). The term "site-specific interactions" used here is a qualitative representation of variety of intermolecular and intramolecular interactions taking place between the atomic-level sites of CO$_2$ molecules and the molecules of liquid in question.

These above-mentioned interesting and useful properties put CO$_2$ into a unique position among supercritical fluids. It is the most widely used supercritical fluid with a wealth of industrial applications and processes such as superfluid extraction (SFE), enhanced oil recovery (EOR), and ScCO$_2$ Brayton power cycles. On the other hand, in the case of EOR, the most efficient use of CO$_2$ (i.e., ScCO$_2$) as an oil-recovery agent (i.e., solvent as well as displacing agent) is obtained at flooding pressures at which miscible displacement is achieved (Holm and Josendal 1974). Hence, achieving miscibility between injected CO$_2$ and reservoir oil is key for any successful EOR project. However, what is the fundamental nature of infinite solubility (i.e., complete miscibility) that occurs between injected CO$_2$ and reservoir oils and is thermodynamically manifested by the MMP (first contact or multiple contacts)? This issue is worth considering and is discussed next.

1.4   Attainment of Miscibility Between Injected CO$_2$ and Reservoir Oil

When used for EOR purposes, CO$_2$ recovers reservoir oil by generating miscibility, swelling the oil, lowering the oil viscosity, and lowering the interfacial tension between the non-contacted reservoir oil and the CO$_2$-oil phase in the near-miscible regions (Lyons and Plisga 2005). Not only can CO$_2$ increase the oil density, but it can also vaporize and extract portions of crude oil (Holm and Josendal 1974). Basically, the physical interactions such as oil swelling, oil density increase, viscosity reduction are the results of a certain process or mechanism that may ultimately result in the

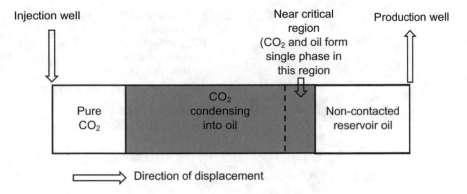

Fig. 1.6 One dimensional schematic of the condensing drive mechanism responsible for the development of CO_2-reservoir oil miscibility in the reservoir, reprinted by permission from Copyright Clearance Center: Springer Nature, Evaluation of CO_2-Oil Miscibility, Saini D, © 2017

development of miscibility in CO_2-reservoir oil mixtures. This process or mechanism is commonly referred to as the condensing drive mechanism in which injected CO_2 condenses into bulk oil phase while displacing it from the injector well to the production well. A schematic description of condensing drive mechanism is shown in Fig. 1.6.

On the other hand, CO_2 can also selectively extract non-polar (apolar) components from the reservoir crude oil (Fang et al. 2016). Hence, in case of certain CO_2-reservoir oil mixtures, the extraction or vaporization of lighter hydrocarbons by injected CO_2 may be the dominating mechanism (i.e., vaporizing drive mechanism), which ultimately results in the development of multiple contact miscibility between injected CO_2 and reservoir oil while injected CO_2 is displacing reservoir oil (Fig. 1.7). Most often, it is a combined or mixed-drive mechanism called a vaporizing/condensing drive mechanism (Zick 1986; Stalkup 1987), which ultimately leads to the development miscibility between injected CO_2 and real reservoir fluids through multiple contacts as injected CO_2 moves downstream while driving the reservoir oil towards the production well. Figure 1.8 schematic depicts this mixed-drive mechanism.

A further understanding of the attainment of miscibility between injected CO_2 and reservoir oil (i.e., a CO_2-reservoir oil system) through multiple contacts, commonly referred as multicontact miscible (MCM) displacement, can be developed on the basis on various theoretical approaches. Some of the sound theoretical approaches are discussed next.

1.4.1   Molecular Dynamics-Based Approach

From a molecular dynamics point of view, miscibility between injected CO_2 and a reservoir oil (i.e., a CO_2-reservoir oil system) can be considered a manifestation of

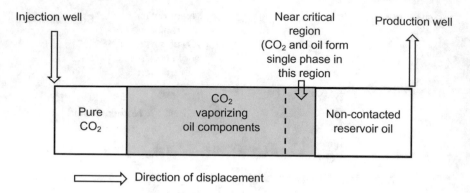

Fig. 1.7 One dimensional schematic of the vaporizing drive mechanism responsible for the development of CO$_2$-reservoir oil miscibility in the reservoir, reprinted by permission from Copyright Clearance Center: Springer Nature, Evaluation of CO$_2$-Oil Miscibility, Saini D, © 2017

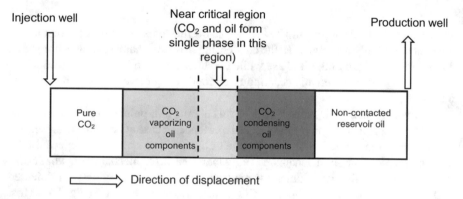

Fig. 1.8 One dimensional schematic of mixed (vaporizing + condensing) drive mechanism responsible for the development of CO$_2$-reservoir oil miscibility in the reservoir, reprinted by permission from Copyright Clearance Center: Springer Nature, Evaluation of CO$_2$-Oil Miscibility, Saini D, © 2017

molecular motion and interaction of molecules (Yang et al. 2016). Both the molecular motion (i.e., thermal motion of molecules due to a system's pressure and temperature conditions) and molecular interactions (i.e., resultant molecular forces arising from intermolecular [van der Waals, electrostatic] and intramolecular [bonding] interactions) dictate the extent of mutual diffusivity (i.e., mass transfer) between injected CO$_2$ and reservoir oil phases and the reduction in CO$_2$-reservoir oil interfacial tension (CO$_2$-reservoir oil IFT). Also, as noted by Orr et al. (1983), molecular size distribution of pentane plus (C$_{5+}$) fraction present in reservoir oil also play a key role in the development of miscibility between injected CO$_2$ and reservoir oil.

The mass transfer is caused by the difference in density or temperature of the two-phase system (Mengshan et al. 2017). For an isothermal process, molecules move from the phase with the larger density to the phase with the smaller density until an equilibrium state is reached (Mengshan et al. 2017). Because CO_2-reservoir oil miscibility is always evaluated at given temperature (i.e., reservoir temperature), density difference is the governing mechanism for the mass transfer via molecular diffusion that takes place between injected CO_2 and reservoir oil phases when they first come in contact of each other.

On the other hand, at lower operating pressures (i.e., injection pressures < MMP), CO_2-reservoir oil mixtures are known to show significant interfacial tension (IFT); however, as pressure approaches near MMP, CO_2-reservoir oil IFT reduces significantly or almost vanishes (e.g., Ayirala and Rao 2006; Sequeira et al. 2008; Saini and Rao 2010). Obviously, the existence of significant IFT between injected CO_2 and a given reservoir oil at lower pressures (< MMP) will result in limited mutual diffusion interactions. However, a significant reduction in IFT at a pressure near MMP facilitates uninterrupted mutual diffusion (i.e., mass transfer) between injected CO_2 and reservoir oil molecules; thus, a homogeneous phase (i.e., no interface) is formed at MMP and the condition of infinite solubility (i.e., complete miscibility) for a given CO_2-reservoir oil system is achieved.

The pressure dependence of CO_2-reservoir oil IFT and its implications for mass transfer can be visualized through Figs. 1.9 and 1.10, which include images taken during the high-pressure high-temperature (HPHT) experimentation conducted by Saini and Rao (2010) for studying the pressure dependence of IFT between pure CO_2 and a real reservoir oil. As can be seen from Fig. 1.9, the regular shape of a freshly introduced pendant drop of reservoir formed with the help of an available injector tip in a HPHT cell, which was initially filled with pure CO_2 and was kept at 20 MPa (2,900 psi) and 142.7 °C (289°F), clearly indicates the presence of significant IFT (2.16 mN/m) between pure CO_2 and fresh reservoir oil. Also, formation of couple of fluid streaks (Fig. 1.9) indicated the presence of week mutual interactions (i.e., limited mass transfer) between freshly-introduced reservoir oil and bulk CO_2 phase.

In contrast, when the pressure is increased to 23.1 MPa (3,500 psi) by injecting fresh CO_2, now a pendant drop of freshly introduced reservoir oil (i.e., reservoir oil of original composition) in the CO_2-rich phase, which is already present in the HPHT cell, could hardly be formed (Fig. 1.10). The system (i.e., CO_2-rich gas phase and oil-rich liquid phase) exhibits a low IFT (0.19 mN/m). Further, the formation of multiple fluid streaks and an immediate distortion of a pendant oil drop (Fig. 1.10) as soon as the reservoir oil (of initial composition) was brought in contact with an existing CO_2-rich phase indicate the presence of strong mutual interactions between freshly-introduced pendant drop of reservoir oil and CO_2-rich phase, which is due to the lack of any significant IFT between the two phases. It is worth mentioning here that the equilibrium compositions of both the phases (i.e., a CO_2-rich gas phase and an oil-rich liquid phase) are significantly different from the original compositions of CO_2 and reservoir oil (i.e., pure CO_2 and reservoir oil of original composition).

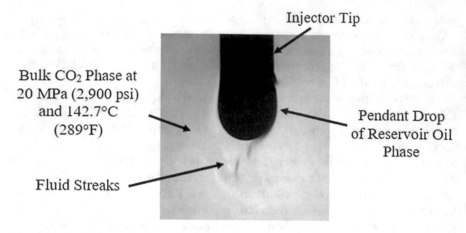

Fig. 1.9 Visualization of weak mutual interactions (i.e., limited mass transfer) between freshly-introduced pendant drop of reservoir oil and bulk CO$_2$ phase at experimental pressure step below MMP and reservoir temperature, as observed during HPHT experimentation conducted by Saini and Rao (2010), modified figure reproduced by permission: Soc Pet Eng, Experimental determination of minimum miscibility pressure (MMP) by gas/oil IFT measurements for a gas injection EOR project. https://doi.org/10.2118/132389-ms, Saini and Rao © 2010

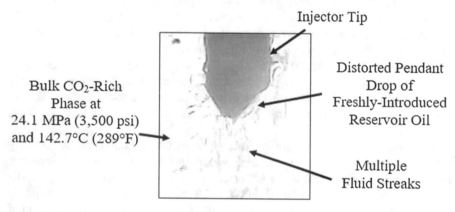

Fig. 1.10 Visualization of strong mutual interaction (i.e., significant mass transfer) between freshly-introduced pendant drop of reservoir oil and CO$_2$-rich phase at experimental pressure step close to MMP and reservoir temperature, as observed during HPHT experimentation conducted by Saini and Rao (2010), modified figure reproduced by permission: Soc Pet Eng, Experimental determination of minimum miscibility pressure (MMP) by gas/oil IFT measurements for a gas injection EOR project. https://doi.org/10.2118/132389-ms, Saini and Rao © 2010

1.4.2 Interfacial Energy-Based Approach

To further understand the delicate interplay of the phenomena of mutual diffusivity and interfacial tension and their implications for the development of CO$_2$-reservoir oil miscibility, we need to look at the concept of interfacial energy (i.e., surface free energy or free surface energy). The theory of the origin of interfacial energy given by Antonoff (1907) in Nosonovsky and Ramachandran (2015) is the simplest way intended to describe the tension at the interface (i.e., IFT) between two liquids (Nosonovsky and Ramachandran 2015). According to Antonow's theory, as explained by Nosonovsky and Ramachandran (2015), the atoms of the surface layer have, on average, fewer bonds with their neighbors; therefore, their energy is larger than that of the atoms or molecules in the bulk (i.e., away from the surface). When two liquid surfaces come into contact, the lack of neighboring bonds in the surface layer is partially compensated by the atoms or molecules in the opposite substance (Nosonovsky and Ramachandran 2015). The total number of interactions are defined by the substance providing the lower numbers of bonds of the two; thus, the interfacial energy is proportional to the difference in the available bonds (Nosonovsky and Ramachandran 2015).

In our case, two liquids will be $ScCO_2$ and a given reservoir oil. Also, it can be expected that reservoir oils, which may possess a wide range of polarities due to the presence of hundreds of different types of hydrocarbon components and various chemical compounds, would have a significantly higher number of available bonds compared to CO_2. Hence, $ScCO_2$, which can act as a weak Lewis acid as well as a Lewis base (i.e., possesses certain polar attributes), is expected to dictate the extent of mutual interactions for a given CO_2-reservoir oil system. It is also noted here that, in the reservoir, due to the simultaneous flow of the injected CO_2 and the displaced reservoir oil, some physical mixing between the two phases always occurs. It is also expected to enhance the physical and chemical interactions that take place when injected CO_2 comes into repeated contacts with reservoir oil present in a confined space (i.e., pore space of the reservoir rock) while moving towards the production well from the injection well.

1.4.3 Thermodynamics-Based Approach

Thermodynamically, for an isothermal and isobaric system, which is the case for the process of CO_2-reservoir oil miscibility, any variation in the chemical potential will tend to lead to mass transport. The chemical potential is actually the Gibbs free energy. the Gibbs free energy represents all the reversible work (such as electric work and work creating surface area) expect mechanical work that can be obtained from the system. The equilibrium IFT, which is a measure of the surface free energy of the system, reflects the condition of the minimization of the Gibbs free energy. The Gibbs free energy must decrease if such a system departs from equilibrium or remains unchanged if an equilibrium has been achieved. Hence, CO_2 and reservoir oil, when encountering each other, attempt to attain an equilibrium by minimizing the Gibbs free energy. In contrast, the Helmholtz free energy, which represents all the reversible work including mechanical work, seeks a minimum, a change in a process occurring at constant temperature and volume.

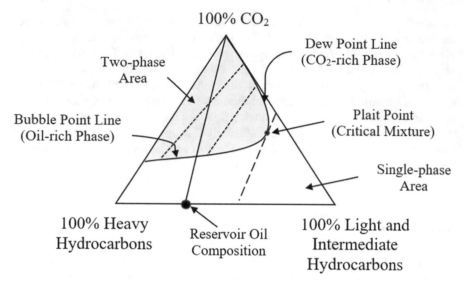

Fig. 1.11 Schematic representation of a multicomponent CO$_2$-reservoir oil mixture as a pseudo ternary system

Nevertheless, CO$_2$-reservoir oil miscibility, whether it is achieved through the process of first-contact or it is achieved through multiple contacts between injected CO$_2$ and reservoir oil, is always described in terms of the thermodynamic state of the system. The thermodynamics state of the system is recognized or identified by the system's pressure (either FC-MMP or MC-MMP [i.e., MMP]) at reservoir temperature.

Even though a CO$_2$-reservoir oil system is a complex multicomponent mixture, it can be simplified to a pseudo ternary system (Fig. 1.11), comprising CO$_2$ and two pseudo hydrocarbon components—namely light and intermediate hydrocarbons and heavy hydrocarbons. As can be seen from Fig. 1.11, the composition of the reservoir oil can be represented by a point on the line joining pseudo hydrocarbon components of the ternary system. Depending on the thermodynamic conditions of imposed pressure and reservoir temperature, CO$_2$ and reservoir oil may form a two-phase mixture, if resultant composition of CO$_2$-reservoir oil mixture is such that it falls within the boundary of two-phase envelope. Outside of the two-phase area, CO$_2$ and reservoir oil can be mixed in any proportion without forming two phases, which will represent the attainment of first-contact miscibility. In other words, if the resultant composition of a CO$_2$-reservoir oil mixture is such that it falls outside the two-phase envelope, first-contact miscibility between CO$_2$ and reservoir oil will occur.

The apex of the two-phase envelope, which is equivalent to the critical point found in traditional thermodynamic phase behavior and is often referred as a plait point, represents the situation where two phases (i.e., a CO$_2$-rich phase and an oil-rich phase) cease to co-exist, and a critical mixture is formed. The system's pressure that corresponds to the achievement of the condition of plait point on a pseudo ternary diagram (e.g., Fig. 1.11) is recognized as the upper critical solution pressure (UCSP) or the FC-MMP.

From a thermodynamic point of view, at the plait point and beyond (i.e., away from two-phase area), it is immaterial if the resulting critical mixture was achieved through the extraction or vaporization of hydrocarbon components by CO_2 (i.e., mixing followed along the dew point line) or CO_2-rich phase condensed into oil-rich phase (i.e., mixing followed along the bubble point line) or a combined effect (i.e., mixing was achieved through vaporizing/condensing mechanism) led to the formation of a critical mixture. In other words, if the phase behavior of a given CO_2-reservoir oil mixture, as a pseudo ternary system, is evaluated at a pressure equal to or higher than UCSP (i.e., FC-MMP), no two-phase mixture will ever form. Hence, CO_2 and reservoir oil, irrespective of their mixing ratio, will be infinitely soluble (i.e., completely miscible).

Thermodynamically, attaining first-contact miscibility does not depend on the mixing ratio of CO_2 and reservoir oil. However, in the pore space of reservoir rock (i.e., pores), after making first ever contact with reservoir oil, both the freshly injected CO_2 and the newly formed CO_2-rich phase continue to move away from the injection well and continue to contact both the fresh reservoir oil and oil-rich phase. These multiple and repetitive contacts between varied combination of different phases (i.e., fresh CO_2, CO_2-rich phase, oil-rich phase, and fresh reservoir oil) along the path of fluid movement (i.e., towards the production well from the injection well) may lead to the development of miscibility at a lower pressure than UCSP or FC-MMP. Hence, the development of miscibility through multiple contacts resulting from the imposed constraints of fluid flow through porous reservoir rock and the phase behavior (i.e., lowest operating pressure and reservoir temperature that results in miscible displacement) is recognized by the MC-MMP (i.e., MMP).

During a miscible flood, injected CO_2 and reservoir oil attain a critical mixture composition through multiple contacts to reduce the residual oil saturation to zero while moving downstream (injection well to production well). Hence, the MMP signifies the thermodynamic condition of lowest operating pressure and reservoir temperature at which CO_2 and reservoir oil develop miscibility through multiple contacts while achieving a local displacement efficiency of 100%. It is important to recognize that it is the combination of flow and phase equilibrium that creates the compositions that lead to high displacement efficiency (Orr and Jessen 2007). The MMP itself is also considered a reservoir-specific parameter because crude oil composition and the reservoir temperature greatly influence its value.

1.5 Other Considerations

It is well recognized that the achievement of miscibility between CO_2 and reservoir oil is critical for miscible CO_2-EOR as well as simultaneous CO_2-EOR and storage projects. However, operating a miscible CO_2 flood at a pressure equal or greater than MMP may not necessarily result in maximized oil recovery. There may be other factors and physical processes which may result in lower recoveries in the field compared to the ones observed in laboratory settings.

For example, both viscous fingering and gravitational fingering (or segregation) can seriously affect the sweep efficiency, thus resulting in low recoveries as found in water alternating gas (WAG) projects involving CO$_2$ injection in miscible mode (Rao et al. 2004). Generally, traditional miscible gas (e.g., CO$_2$, hydrocarbon gas) EOR projects are very sensitive to geological heterogeneity (Muggeridge et al. 2014), and CO$_2$-EOR recovery factor generally shows a dependence to injected volume (Olea 2017). The precipitation of asphaltenes, even in low amounts, may hinder fluid flow and significantly reduces the mobility of injected CO$_2$ (Perera et al. 2016).

The role of the above-mentioned factors and physical processes in influencing the performance of miscible CO$_2$ flooding needs to be thoroughly examined while planning, designing, and implementing EOR projects. Nevertheless, for a given CO$_2$ and reservoir oil system, the determination of CO$_2$-reservoir oil miscibility in terms of the MMP and the characterization of the drive mechanism(s) that are responsible for the attainment of CO$_2$-reservoir oil miscibility serves as a guide to decide the operating pressure at which an effective miscible displacement, and thus high oil recovery, can be achieved.

References

Antonoff GN (1907) Sur La Tension Superficielle Des Solutions Dans La Zone Critique. J Chim Phys 5:364–371

Ayirala SC, Rao DN (2006) Comparative evaluation of a new MMP determination technique. Soc Pet Eng. https://doi.org/10.2118/99606-MS

Bruening IMRA (1991) Crude oil polarity measures quality, predicts behavior. Oil Gas J 89(31). http://www.ogj.com/articles/print/volume-89/issue-31/in-this-issue/refining/crude-oil-p olarity-measures-quality-predicts-behavior.html. Accessed 31 Dec 2017

Burke J (1984) Solubility parameters: theory and application. In: The book and paper group annual 3. The American Institute for Conservation. https://cool.conservation-us.org/coolaic/sg/bpg/ann ual/v03/bp03-04.html. Accessed 30 Dec 2017

Clark NJ, Shearin HM, Schultz WP et al (1958) Miscible drive—its theory and application. Soc Pet Eng. https://doi.org/10.2118/1036-G

Fang T, Shi J, Sun X et al (2016) Supercritical CO$_2$ selective extraction inducing wettability alteration of oil reservoir. J Supercrit Fluids 113:10–15

Global CCS Institute (2011) Bridging the commercial gap for carbon capture and storage. http://www.globalccsinstitute.com/publications/bridging-commercial-gap-carbon-capture-and-storage. Accessed 25 Dec 2017

Holm LW, Josendal VA (1974) Mechanisms of oil displacement by carbon dioxide. Soc Pet Eng. https://doi.org/10.2118/4736-PA

Imre A, Ramboz C, Kraska T et al (2015) Anomalous fluid properties of carbon dioxide in the supercritical region—Application to geological CO$_2$ storage and related hazards. Environ Earth Sci Springer 73:4373–4384

Kaziunas A, Schlake R (2016) Why is CO$_2$ a green solvent. ACS Green Chemistry Institute. https://communities.acs.org/community/science/sustainability/green-chemistry-nexus-blog/blog/2016/06/10/why-is-co2-a-green-solvent. Accessed 29 Dec 2017

Lyons WC, Plisga GJ (2005) Standard handbook of petroleum and natural gas engineering. Gulf Professional Publishing, Houston

Mengshan L, Liang L, Xingyuan H et al (2017) Prediction of supercritical carbon dioxide solubility in polymers based on hybrid artificial intelligence method integrated with the diffusion theory. RSC Adv 7:49817–49827

Meyer JP (2007) Summary of carbon dioxide enhanced oil recovery (CO_2 EOR) injection well technology. American Petroleum Institute. http://www.api.org/~/media/files/ehs/climate-chang e/summary-carbon-dioxide-enhanced-oil-recovery-well-tech.pdf. Accessed 22 Jan 2017

Muggeridge A, Cockin A, Webb K et al (2014) Recovery rates, enhanced oil recovery and techno-logical limits. Phil Trans R Soc A:372. https://doi.org/10.1098/rsta.2012.0320

Net Industries (2017) Miscibility. http://science.jrank.org/pages/4382/Miscibility.html. Accessed 26 Dec 2017

NETL (2010) Carbon dioxide enhanced oil recovery: untapped domestic energy supply and long term carbon storage solution. The U.S. Department of Energy's (DOE) National Energy Tech-nology Laboratory. https://www.netl.doe.gov/file%20library/research/oil-gas/CO2_EOR_Prime r.pdf. Accessed 26 Dec 2017

Nosonovsky M, Ramachandran R (2015) Geometric interpretation of surface tension equilibrium in superhydrophobic systems. Entropy 17:4684–4700. https://doi.org/10.3390/e17074684

Olea RA (2017) Carbon dioxide enhanced oil recovery performance according to the literature. In: Verma MK (ed) Three approaches for estimating recovery factors in carbon dioxide enhanced oil recovery, U.S. Geological Survey. Scientific Investigations Report–5062: D1–D21

OpenStax (2016) Solubility. In: Chemistry. Rice University. https://cnx.org/contents/havxkyvS@ 9.311:2488fW6W@5/Solubility. Accessed 27 May 2018

Orr FM, Jessen K (2007) An analysis of the vanishing interfacial tension technique for determination of minimum miscibility pressure. Fluid Phase Equilib 255(2):99–109

Orr FM, Silva MK, Lien CL (1983) Equilibrium phase compositions of CO_2/crude oil mixtures-part 2: comparison of continuous multiple-contact and slim-tube displacement tests. Soc Pet Eng. https://doi.org/10.2118/10725-PA

Peach J, Eastoe J (2014) Supercritical carbon dioxide: a solvent like no other. Beilstein J Org Chem 10:1878–1895. https://doi.org/10.3762/bjoc.10.196

Perera MSA, Gamage RP, Rathnaweera TD et al (2016) Review of CO_2-enhanced oil recovery with a simulated sensitivity analysis. Energies 9(481)

Qin J, Han H, Liu X (2015) Application and enlightenment of carbon dioxide flooding in the United States of America, Petrol Explor Develop 42 (2):232-240

Rao DN, Ayirala SC, Kulkarni MM et al (2004) Development of Gas Assisted Gravity Drainage (GAGD) process for improved light oil recovery. Soc Pet Eng. https://doi.org/10.2118/89357-MS

Raveendran P, Ikushima Y, Wallen SL (2005) Polar attributes of supercritical carbon dioxide. Accou Chem Res 38(6):478–485

Saini D (2017) Engineering aspects of geologic CO_2 storage: synergy between enhanced oil recovery and storage. Springer International Publishing AG, Switzerland

Saini D, Rao DN (2010) Experimental determination of minimum miscibility pressure (MMP) by gas/oil IFT measurements for a gas injection EOR project. Soc Pet Eng. https://doi.org/10.211 8/132389-MS

Sequeira DS, Ayirala SC, Rao DN (2008) Reservoir condition measurements of compositional effects on gas-oil interfacial tension and miscibility. Soc Pet Eng. https://doi.org/10.2118/1133 33-MS

Span R, Wagner W (1996) A new equation of state for carbon dioxide covering the fluid region from the triple-point temperature to 1100 K at pressures up to 800 MPa. J Phys Chem Ref Data 25(6):1509–1596

Stalkup FI (1987) Displacement behavior of the condensing/vaporizing gas drive process. Soc Pet Eng. https://doi.org/10.2118/16715-MS

Suico J (2017) What are the differences between solubility and miscibility? https://sciencing.com/ differences-between-solubility-miscibility-8663853.html. Accessed 26 Dec 2017

Supercriticalfluid website (2017) Explore, use, make the most of supercritical fluids. http://www.supercriticalfluid.org/Supercritical-fluids.146.0.html. Accessed 29 Dec 2017

Tyrer D (1912) The theory of solubility. J Phys Chem 16(1):69–85

Whorton LP, Brownscombe ER, Dyes AB (1952) Method for producing oil by means of carbon dioxide. US Patent 623,596, 30 Dec 1952

Yang S, Lian L, Yang Y et al (2016) Molecular dynamics simulation of miscible process in co2 and crude oil system. Soc Pet Eng. https://doi.org/10.2118/182907-MS

Zick AA (1986) A combined condensing/vaporizing mechanism in the displacement of oil by enriched gases. Soc Pet Eng. https://doi.org/10.2118/15493-MS

Chapter 2
Characterization and Determination of CO_2-Reservoir Oil Miscibility

Abstract For assisting in making an informed decision about the selection of an experimental or a non-experimental approach to characterize and determine CO_2-reservoir oil miscibility in terms of the MMP, a quick review of the needs to determine the MMP and various factors including the technical and economic factors that might have a great influence on such decision are presented and discussed in detail.

2.1 Needs to Determine the MMP

The main premise behind operating a CO_2 injection-based EOR project in miscible mode is that it offers the highest oil recovery potential. In their patent for the method for producing oil by means of CO_2, Whorton et al. (1952) clearly demonstrated that highest oil recovery was achieved when 100% CO_2 was injected at a certain pressure. It has been long recognized that, by using a miscible solvent as a displacing agent, the oil in a reservoir can be displaced from the rock with a high degree of efficiency (Clark et al. 1958). Also, if the solvent is CO_2, then the minimum pressure required for achieving miscible displacement (i.e., MMP) is usually significantly lower than the pressure required for miscible displacement with either natural gas, flue gas, or nitrogen (Stalkup 1978).

The achievement of miscible displacement between displacing fluid (e.g., CO_2) and displaced fluid (e.g., oil) is mainly for achieving maximum reduction in residual oil saturation. The extent of reduction in residual oil saturation (i.e., the oil left behind in the pore space after being contacted with injected CO_2) is the key difference between the miscible and immiscible displacement processes. In miscible displacements, the residual oil saturation is reduced nearly to zero, which leads to high oil recoveries and favorable project economics, whereas considerable residual oil saturation can remain in immiscible displacements, which often results in unfavorable project economics (Meyer 2007).

As mentioned earlier, majority of the CO_2 injection-based EOR projects around the world are miscible flooding projects, which itself is evidence of the technical as well as the economic success of CO_2 injection-based miscible displacement pro-

D. Saini, *CO2-Reservoir Oil Miscibility*, SpringerBriefs in Petroleum Geoscience & Engineering, https://doi.org/10.1007/978-3-319-95546-9_2

cesses. Also, the conventional water alternating gas (WAG) technique is the most popular form of the operating method, used in over 90% of the miscible CO_2-EOR projects implemented to date (Merchant 2017). In the WAG technique, CO_2 and water are alternatively injected into the reservoir. The size, expressed in terms of a fraction of the reservoir's hydrocarbon pore volume (HCPV) and injection durations of both CO_2 and water slug cycles, varies from project to project.

The use of CO_2 flooding technology (miscible or immiscible) is mainly centered in the United States as evident from the United States' contribution (over 90%) to the world CO_2-EOR oil production (Qin et al. 2015). The United States also leads the world in using immiscible CO_2 flooding techniques. Almost half (46%) of such projects worldwide are in the United States (Zhang et al. 2018), however, immiscible CO_2 flooding only accounts for less than 10% of yearly CO_2-EOR oil production (Qin et al. 2015). One of the operating companies—Occidental Petroleum Corporation (OXY)—alone accounts for 36% of the US annual miscible CO_2 flooding oil production (Qin et al. 2015). Other major US CO_2 miscible flooding operators include Kinder Morgan, Chevron, Hess, and Denbury Resources. According to Qin et al. (2015), the contribution of these companies to the US annual miscible CO_2 flooding oil production ranges from 6.9% (Denbury Resources) to Kinder Morgan (11%).

Other operating methods include huff-and-puff, gravity drainage, residual oil zone (ROZ), double displacement, gas cycling, horizontal well pattern development, and CO_2 gas drive with nitro boost (Merchant 2017). The viability of miscible CO_2 flooding is being tested or have already been evaluated for major U.S. shale plays like the Bakken (Jin et al. 2017) and the Eagle Ford (Gamadi et al. 2014). Because miscibility can hardly be reached in heavy oil reservoirs, CO_2 flooding in immiscible mode has also been tested in heavy oil reservoirs like Wilmington field, California (Saner and Patton 1986).

Irrespective of type of operating method, if miscibility is desired in a CO_2-EOR or simultaneous CO_2-EOR and storage project, a reliable estimation of CO_2-reservoir oil MMP is always sought. The CO_2-reservoir oil MMP, which is simply referred as the MMP throughout the text, is also one of the primary screening criteria that a reservoir must meet before its suitability for a CO_2-EOR project is further evaluated. As emphasized by Khazam et al. (2016), the MMP is the single most important parameter in the design of a miscible flood because a reliable estimation of the MMP helps the operator to develop injection conditions and to plan suitable surface facilities. On the other hand, based on many laboratory tests and some field implementations, it has been suggested that a degree of immiscibility (i.e., keeping operating pressure in the near-miscible region), characterized by the name "near miscible" is sufficient, because near-miscible and miscible gases perform in a comparable manner (Thomas et al. 1994; Dong et al. 2001). However, as name suggests, determination of "near miscible" region itself relies on the availability of information on the MMP for given CO_2-reservoir oil system.

According to Olden et al. (2015), maintaining the reservoir pressure high enough to ensure CO_2-reservoir oil miscibility is the key for achieving higher oil recovery factors. On the other hand, a presence of varied permeability in the reservoir (i.e.,

Table 2.1 Various experimental approaches reported in the published literature for characterizing and determining CO_2-reservoir oil miscibility in terms of the MMP (modified by permission within the scope of open access article from Saini (2016a), data source include Alomair and Garrouch (2016))

Experimental approaches	
Method name	Literature source
Slim-tube method	Yellig and Metcalfe (1980)
High-pressure visual sapphire cell	Hagen and Kossak (1986)
Rising-bubble apparatus method	Christiansen and Haines (1987)
Vapor density of injected gas versus pressure	Harmon and Grigg (1988)
PVT multi-contact experiments	Thomas et al. (1994)
Vanishing interfacial tension (VIT) method	Rao (1997)
Single bubble injection technique	Srivastava and Huang (1998)
Vapor liquid equilibrium-interfacial tension test	Kechut et al. (1999)
Micro slim-tube test	Kechut et al. (1999)
Fast fluorescence-based microfluidic method	Nguyen et al. (2015)
Fast slim-tube method	Adel et al. (2016)
Capillary-rise-based VIT method	Hawthorne et al. (2016)
Sonic response method	Czarnota et al. (2017)
Rapid pressure increase method	Czarnota et al. (2017a)

a heterogeneous reservoir) can lead to a heterogeneous pressure distribution in the reservoir leading to heterogeneous miscibility. In other words, as the CO_2 injection progresses in the reservoir, over time, some miscible zones may turn into immiscible zones due to either change in residual oil composition or due to the presence of heterogeneous pressure distribution (Changlin et al. 2016). Hence, it is necessary that both the phase behavior and the flow aspects are considered while characterizing and determining CO_2-reservoir oil miscibility.

The phenomenon of CO_2-reservoir oil miscibility can either be characterized and determined experimentally or it can also be characterized and determined using a non-experimental (empirical, analytical, and computational/numerical) approach. Tables 2.1 and 2.2 list different experimental and non-experimental (empirical, analytical, and computational or numerical) MMP determination methods that have been reported in the published literature. Some of these methods are discussed in detail in later chapters (Chaps. 3 and 4). However, a brief discussion of the key technical factors and economic reasons, which call for the determination of the MMP as the next evaluation step after initial screening of candidate reservoirs, is provided next.

Table 2.2 Various non-experimental approaches reported in the published literature for characterizing and determining CO_2-reservoir oil miscibility in terms of the MMP (modified with permission within the scope of open access article from Saini (2016a), data sources include Mogensen et al. (2009); Alomair and Garrouch (2016); and Zhang et al. (2015))

Non-experimental approaches	
Method name	Literature source
Empirical approach	
Empirical correlations	Cronquist ((1977) in Mungan (1981)), Lee (1979), Yellig and Metcalfe (1980), Johnson and Pollin (1981), Orr and Jensen (1984), Alston et al. (1985), Glaso (1985), Emera and Sarma (2005), Yuan et al. (2005), Shokir (2007), Ghomian et al. (2008), Li et al. (2012), Ju et al. (2012), Chen et al. (2013), Alomair and Iqbal (2014), Alomair and Garrouch (2016), Zhang et al. (2015), Valluri et al. (2017), Mansour et al. (2017)
Analytical approach	
Key Tie-Line approach and the Method of Characteristics (MOC)	Nouar and Flock (1988), Johns et al. (1993), Johns and Orr (1996), Wang and Orr (1997), Jessen et al. (1998), Yuan and Johns (2005), Ahmadi et al. (2011), Yuan et al. (2008)
Mechanistic Parachor models	Ayirala and Rao (2004), Ayirala and Rao (2006), Ashrafizadeh and Ghasrodashti (2011)
Computational or numerical approach	
Mixing-cell methods	Jaubert et al. (1998, 1998a), Zhao et al. (2006, 2006a), Ahmadi and Johns (2011), Mogensen et al. (2009)
1D slimtube simulation models	Mogensen et al. (2009), Bui et al. (2010), Ju et al. (2012)
Linear gradient theory (LGT) model	Nobakht et al. (2008)
Neural network model	Alomair and Garrouch (2016)
Least-squares support vector machine (LSSVM) model	Shokrollahi et al. (2013)
Parachor models	Teklu et al. (2014)
PC-SAFT EOS model	Wang et al. (2016)
Diminishing interface method	Zhang et al. (2017)
Molecular dynamic simulation	Yang et al. (2016)

2.2 Technical Factors

The main technical reason behind the determination of CO_2-reservoir oil miscibility in terms of the MMP and characterization of dominating drive mechanism(s) for a given reservoir oil and the CO_2 injection stream is that they are greatly influenced by the reservoir temperature, the composition and certain PVT properties of the reservoir oil, and the composition of the injected CO_2. Hence, they need to be determined for project-specific CO_2-injection stream (pure or impure) and reservoir oil system. Obviously, from a technical point of view, we need to consider certain thermodynamic variables, influence of natural forces, and operational factors that are directly correlated to the achievement of miscibility and maximum oil recovery.

2.2.1 Thermodynamic Variables

Based on the comprehensive analysis of 113 MMP measurements (pure CO_2) taken from worldwide gas injection projects published in the literature, Alomair et al. (2015) studied the influence of various independent variables on the MMP. According to them, reservoir temperature is the most influential factor in determining the MMP. Generally, it is observed that MMP increases with reservoir temperature.

Various researchers (e.g., Holmes and Josendal 1974; Yellig and Metcalfe 1980; Alston et al. 1985; Emera and Sarma 2005; Shokir 2007; Chen et al. 2013) have documented the fact that, apart from reservoir temperature, achievement of miscibility between injected CO_2 and reservoir oil is strongly correlated to oil composition and molecular weight of heavier hydrocarbon fraction (i.e., pentane plus [C_{5+}] or heptane plus [C_{7+}]) present in the reservoir oil. Certain phase behavior properties of reservoir oil, especially the API gravity, bubble point pressure, and initial dissolved gas/oil ratio, also play an important role in determining the miscibility between injected CO_2 and a given reservoir oil. However, as demonstrated in various published studies (e.g., Rezaei et al. 2013; Alomair et al. 2015; Karkevandi-Talkhooncheh et al. 2018), ratio of volatile (C_1 and N_2) to intermediate (C_2–C_6, CO_2, and H_2S) has a major effect on the development of CO_2-reservoir oil miscibility.

On other hand, for a variety of reasons, pure CO_2 is not always available for CO_2-EOR and/or simultaneous CO_2-EOR and storage projects. In the case of CO_2-EOR projects, recycled CO_2 (i.e., amount of CO_2 that is produced back to the surface during oil production) is injected back, which helps in keeping the purchased volumes of new CO_2 at minimum level. Also, the use of anthropogenic CO_2 as well as the necessity to inject the recycled CO_2 back for maximizing both the oil recovery and the CO_2 storage in simultaneous CO_2-EOR and storage projects are the leading causes behind the introduction of impurities in the injected CO_2 stream. A typical impure CO_2 stream may contain a variety of impurities including nitrogen (N_2), hydrogen sulfide (H_2S), and lighter hydrocarbon components.

Glaso (1985) reported that even small impurities can significantly affect the miscibility pressure. The addition of H_2S and C_2 +hydrocarbons to an injected CO_2 stream is known to lower the miscibility pressure, whereas the pressure of C_1 in the injected CO_2 stream results an increase in the miscibility pressure (Metcalfe 1982). According to Sebastian et al. (1985), the detrimental effect of light gases such as N_2 and methane can be balanced against the beneficial effects of intermediate hydrocarbon components (C_2–C_6) to maintain a relatively stable MMP. Such balance in relative compositions of various impurities in injection stream largely depends on the resultant weight fraction average critical temperature or mole fraction average critical temperature of the injected impure CO_2 stream. Though the average weight or mole fraction average critical temperature of a given impure CO_2 stream can be determined easily, the determination of the MMP of a given impure CO_2 stream and reservoir oil system still needs to be performed either experimentally or theoretically (i.e., using an analytical or a computational/numerical method).

2.2.2 Natural Forces

Apart from the above-mentioned thermodynamic factors, there are certain natural forces—namely viscous, capillary, and gravity forces—that have a great influence on the oil recovery from CO_2-EOR projects. Most often, the effects of these natural forces are quantified in terms of dimensionless numbers. In case of miscible flooding, the capillary number, which signifies the relative importance of viscous and capillary forces, plays a critical role in determining residual oil saturation thus, highlighting the need to characterize and determine the MMP.

2.2.2.1 Capillary Number

The capillary number, which is generally denoted as Ca or N_{ca}, is the manifestation of complex interactions between viscous and capillary forces and reservoir wettability, thus determining the residual oil saturation in a reservoir that can be achieved by a displacement process devised for the EOR purpose. A generalized form of a capillary number can be expressed as:

$$Ca\, or\, N_{ca} = \frac{v_{CO_2}\mu_{CO_2}}{\phi\sigma_{CO_2-oil}cos\theta} \tag{2.1}$$

where v_{CO_2} is the interstitial or linear velocity of the injected CO_2 in porous reservoir rock (derived from the Darcy's Law) of porosity ϕ, μ_{CO_2} is the viscosity of injected CO_2 at a given operating pressure, and σ is the IFT between the displaced (oil) and displacing (CO_2) fluid. θ is the contact angle which is a measure of wetting preference (i.e., water wet, oil-wet, intermediate wet, or mixed wet) of the reservoir rock.

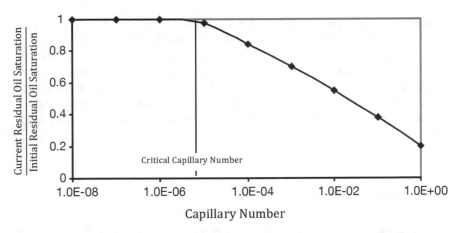

Fig. 2.1 Typical capillary desaturation (i.e., normalized residual oil saturation [current residual oil saturation/initial residual oil saturation] versus capillary pressure) curve. Modified by permission: Soc Pet Eng, New approach in gas injection miscible processes modelling in compositional simulation: Petroleum Society of Canada, Shtepani E, F.B. Thomas FB, Bennion DB © (2006)

A typical relationship between the capillary number and the normalized residual oil saturation (i.e., residual oil saturation/initial oil saturation), which is typically shown through a capillary desaturation curve, is depicted in Fig. 2.1. As can be seen in Fig. 2.1, a significant reduction in residual oil saturation can be achieved by lowering the capillary number. However, it requires orders of magnitude changes in the capillary number before significant decreases in residual oil saturation could be achieved (Thomas et al. 1994). On the other hand, by definition, miscibility itself is the condition of no interface (i.e., zero CO_2-reservoir oil IFT). As evident from Eq. 2.1, a condition of zero IFT between reservoir oil and injected CO_2 will result in an infinite capillary number, which implies that no oil is left behind in the CO_2-contacted portion of the reservoir.

In most reservoirs, a near-MMP operating pressure, which signifies the achievement of near-miscible conditions and a significant reduction in IFT, is more than sufficient for reduced residual oil saturation and more efficient sweep (Thomas et al. 1994). On another hand, if reservoir wettability is not favorable (i.e., reservoir is preferentially oil-wet) or the tighter pore throats, which are essentially isolated from the injected fluid at a higher IFT, achievement of zero IFT (i.e., achievement of miscible condition) would be necessary (Thomas et al. 1994). The influence of reservoir wettability (quantified in terms of contact angle, θ), pore size (quantified in terms of radius of pore throat, r), and IFT (σ) on the extent of capillary forces can be understood from the Laplace Equation (Eq. 2.2), which is used to study the role of capillary forces (in term of capillary pressure, P_c) on the behaviors of moving fluid phases in porous medium.

$$P_c = \frac{2\sigma \cos\theta}{r} \qquad (2.2)$$

A considerable decrease in IFT, which is a thermodynamic quantity like the MMP, is relatively easy to achieve and less costly, because the achievement of even one order of magnitude increase in capillary number by increasing velocity or viscosity is much more difficult and costly (Thomas et al. 1994). Hence, it is imperative that reliable information on the MMP for the CO_2-reservoir oil system in question is available if the resulting influence of two leading natural forces (i.e., viscous forces and capillary forces) on oil recovery needs to be minimized in a cost-effective and efficient manner.

2.2.3 Operational Factors

Certain operational factors also signify the need to determine and characterize the MMP. During the analysis of a comprehensive dataset collected from 134 CO_2-EOR projects in the USA, Yin (2015) found that, in general, the CO_2 injection pressure is kept about 1.4 MPa (200 psia) higher that the MMP to ensure that miscibility can be achieved, or to keep reservoir pressure between the MMP and the fracture pressure (Yin 2015). It is noted here that reported MMP data were available for only 22 projects.

While evaluating the CO_2-EOR and storage potential of candidate Californian reservoirs, Saini (2015) illustrated that maintaining the reservoir operating (i.e., injection) higher than the MMP is beneficial from a storage point of view; however, it does not improve the oil recovery further. As can be seen in Tables 2.3 and 2.4, for both the Reservoirs X and Y evaluated by Saini (2015), results of sensitivity analysis performed to assess the influence of operating pressure to MMP ratio on expected oil recoveries revealed that oil recoveries were largely insensitive to operating pressure increase beyond the MMP. No significant change in oil recoveries was observed when the operating pressure to MMP ratio was varied from an initial value of 1.1 to a maximum value of 1.6.

The knowledge of MMP also helps in determining the additional steps that might be necessary for achieving miscibility in the reservoir, if reservoir pressure is lower than the MMP. These additional steps may include the re-pressurization of the reservoir via water injection or CO2 injection itself and/or addition of other gases to lower the MMP.

2.3 Economic Factors

From an economic point of view, the lower the MMP, the lower would be the maintenance and injection costs for operating the project. A prior knowledge of the MMP also assists the operator in maintaining the pressure of both the injection zone and overlying cap rock within maximum allowable limits, which are generally governed by their fracture pressures. Obviously, a low operating or injection pressure is always

Table 2.3 Field wide CO_2-EOR and storage potential predication results for Reservoir X at 1 hydrocarbon pore volume (HCPV) fluid injection (modified by permission from Saini (2015))

Ratio of operating reservoir pressure to MMP	Ratio of operating reservoir pressure to fracture pressure	Injection mode	Field injection rate (HCPV/Year)	Field CO_2 injection rate (Million Metric Tons/Year)	Injection duration (Years)	Incremental oil recovery (%)	Total amount of CO_2 stored (Million Metric Tons)
1.10	0.51	Continuous miscible CO_2	0.03	1.00	34	7.4	6.35
1.40	0.63	Continuous miscible CO_2	0.03	1.20	34	7.8	7.80
1.60	0.72	Continuous miscible CO_2	0.03	1.25	34	8.0	8.60
1.10	0.51	1:1 WAG	0.03	0.50	34	19.8	8.90
1.40	0.63	1:1 WAG	0.03	0.60	34	19.8	10.55
1.60	0.72	1:1 WAG	0.03	0.625	34	19.8	11.35

Table 2.4 Field wide CO_2-EOR and storage potential predication results for Reservoir Y at 1 hydrocarbon pore volume (HCPV) fluid injection (modified by permission from Saini (2015))

Ratio of operating reservoir pressure to MMP	Ratio of operating reservoir pressure to fracture pressure	Injection mode	Field injection rate (HCPV/Year)	Field CO_2 injection rate (Million Metric Tons/Year)	Injection duration (Years)	Incremental oil recovery (%)	Total amount of CO_2 stored (Million Metric Tons)
1.10	0.47	Continuous miscible CO_2	0.04	0.90	25	18.7	7.20
1.40	0.59	Continuous miscible CO_2	0.04	1.08	25	19.7	9.10
1.60	0.68	Continuous miscible CO_2	0.04	1.10	25	20.3	10.10
1.10	0.47	1:1 WAG	0.04	0.45	25	38.0	8.55
1.40	0.59	1:1 WAG	0.04	0.54	25	37.8	10.32
1.60	0.68	1:1 WAG	0.04	0.59	25	37.9	11.25

a plus point when it comes to maintain the project profitable especially in low oil price regime. The previous and current onshore experiences, mainly in the Permian Basin region of Texas, USA, where a majority of the CO_2-EOR projects are located, have shown that overall project economics are good if a low-cost CO_2 source such as natural CO_2 deposits can be used (Gozalpour et al. 2005).

CO_2-EOR projects are primarily designed to maximize the oil recovery and to minimize the amounts of new (i.e., purchased) CO_2, since CO_2 costs money to transport and inject (Qi at el. 2008). Miscible CO_2 floods are found to be economically attractive in many reservoirs (Lorsong 2013). That is why, a majority of CO_2-EOR projects around the world rely on miscible flooding techniques. On the other hand, a renewed interest in economically viable geologic CO_2 storage component coupled with CO_2-EOR (i.e., simultaneous CO_2-EOR and storage project) has prompted several operators to convert certain existing CO_2-EOR projects into simultaneous CO_2-EOR and storage projects and to design and implement new CO_2-EOR projects as simultaneous CO_2-EOR and storage projects (Saini 2017). It is worth mentioning here that a difference between pure CO_2-EOR and simultaneous CO_2-EOR and storage project arises from the fact that the operator needs to take certain additional steps to further monitor, account for, and verify the safe and long-term storage of the injected CO_2 in a simultaneous CO_2-EOR and storage project, whereas, additional monitoring, accounting, and verification of incidentally stored CO_2 in the reservoir is not done for a pure CO_2-EOR project.

Saini (2017) has provided a detailed discussion on the monitoring, verification, and accounting (MVA) or measurement, monitoring, and verification (MMV) program that transforms a commercial CO_2-EOR project into a simultaneous CO_2-EOR and storage project.

Based on the sensitivity analysis performed during the engineering-economic modeling of four case study reservoirs, where simultaneous CO_2-EOR and storage projects either currently operating or were operated in the past, McCoy and Rubin (2009) concluded that the breakeven CO_2 price (i.e., maximum cost of new [i.e., purchased] CO_2 that can be afforded by an operator to operate a simultaneous CO_2-EOR and storage project without incurring a loss) is highly sensitive to several factors. The factors include oil price escalation rate, loss of injected CO_2 in the reservoir (i.e., incidental storage), reservoir pressure, temperature, and initial oil saturation (i.e., residual to waterflooding).

Interestingly, the influence of all the above-mentioned factors except one (oil price) on project economics can also be collectively studied in terms of the ratio between the operating reservoir pressure to the MMP as demonstrated by Saini (2015). It can be seen that, in Tables 2.3 and 2.4, as the operating reservoir pressure to MMP ratio increases, the amount of total CO_2 stored in reservoir (million metric tons, MMt) also increases. However, higher ratios (>1.1) have a negligible effect on incremental oil recovery. It is imperative that operation of a miscible flood at a reservoir pressure significantly higher than the MMP would result in a maximum cost of new (i.e., purchased) CO_2 without any significant gain in incremental oil recovery.

On the other hand, results shown in Tables 2.3 and 2.4 also indicate that the WAG method, compared to continuous CO_2 injection, results in both significantly higher oil recoveries and the amounts of CO_2 stored even though the objective of WAG method is to minimize the amount of CO_2 purchased (i.e., the amount of CO_2 that is incidentally stored [left-behind] in the reservoir). Hence, a prior knowledge of the MMP for a given CO_2-reservoir oil system can also help in optimizing the overall project economics for a given project type (e.g., pure CO_2-EOR or simultaneous CO_2-EOR and storage) employing a given operating method (e.g., WAG or continuous CO_2 injection).

2.4 Is It Necessary to Characterize and Determine CO_2-Reservoir Oil Miscibility?

After considering various technical and economic factors, establishing that the MMP is indeed a key variable needs to be known in the planning and designing phase of a miscible CO_2 flooding (EOR or simultaneous EOR and storage) project. As emphasized by Jaubert et al. (1998a, b), like any other equilibrium thermodynamic property of a system, the MMP is not influenced by the relative permeability, capillary pressure, and interfacial tension. Hence, highest possible oil recoveries can be expected, if reservoir pressure is always maintained above the MMP. On the other hand, as emphasized by Johns et al. (2010), understanding the impact of the gas compositional changes on the MMP is essential to optimize the design of field-wide pressure management and utilization of injected CO_2. Hence, it is important that CO_2-reservoir oil miscibility in terms of the MMP is characterized and determined accurately and reliably. It can be either be done experimentally or theoretically. However, certain experimental methods—namely, slim tube tests—are considered expensive and time consuming, operators tend to rely either on less expensive and fast experimental approaches for obtaining the desired information or they use non-experimental approaches such as an analytical or a computational/numerical approach for calculating the MMP.

On the other hand, there are certain published studies (e.g., Jaubert et al. 2002) that have concluded that, when the injected gas is not pure CO_2 (or N_2 or H_2S), it is not necessary to measure the MMP. Instead, it is enough to fit only two parameters of the equation of state (EOS) data including classical phase behavior (PVT data), swelling behavior (swelling data), and multi-contact behavior (MCT data) and then to predict the MMP using the tuned equation of state model. However, MMP data can also be used directly in regression for further tuning of an EOS model for using the EOS model for performing compositional reservoir studies aimed to evaluate the future performance of a CO_2-EOR or a simultaneous CO_2-EOR and storage projects (e.g., Saini et al. 2013).

Nevertheless, a basic understanding of both the experimental and non-experimental (empirical, analytical, and numerical) approaches, which are currently available for determining and characterizing CO_2-reservoir oil miscibility in terms of the MMP, is always desired for selecting a suitable method that is best suited for serving the project's needs. For example, in actual field application, use of pure CO_2 may prove to be impractical from economic point of view as removal of tract impurity gases from CO_2 injection stream may be cost prohibitive. In such scenario, characterization and determination of CO_2-reservoir oil miscibility in terms of the MMP for impure CO_2-reservoir oil system by either an experimental or a non-experimental approach is a necessity.

2.5 CO_2-Reservoir Oil Miscibility Characterization and Determination Approaches

Different experimental and non-experimental approaches (Table 2.1) have been reported in the published literature for characterizing and determining the CO_2-reservoir oil miscibility in terms of the MMP for a given CO_2-EOR and/or a simultaneous CO_2-EOR and storage project. However, it is the nature of the project which justifies the selection of a given approach. For example, if large numbers of candidate reservoirs need to be screened for determining if CO_2-reservoir oil miscibility is achievable in them (i.e., what is expected MMP value), empirical correlations are best suited for providing the desired information in a rapid and cost-effective manner. On another hand, if quality fluid phase behavior (i.e., PVT) data are available for developing a well-tuned EOS model, then the operator can opt for a computational (numerical) or analytical approach for determining the initial estimates of the MMP for a candidate reservoir.

In the case of a field-scale CO_2-EOR or simultaneous CO_2-EOR and storage project, where further tuning of compositional dynamic simulation model(s) is needed for better history matching and better prediction of future field performance, the operator may decide to experimentally determine the MMP for its CO_2-reservoir oil system for reconfirming the MMP estimates made from non-experimental approaches. The experimentally determined MMP data can also help in further calibration of the compositional dynamic simulation model(s) and thus increase the confidence in the simulation results.

The slim-tube test, which is considered an industry standard for obtaining immediate information about potential operating pressures for a given CO_2-EOR or a simultaneous CO_2-EOR and storage project, is a widely accepted and commonly used experimental technique for measuring the MMP. Other frequently-used experimental techniques include the rising bubble apparatus (RBA) and the vanishing interfacial tension (VIT) technique. Recently, the use of fluorescence-based microfluidic method has also been reported in the published literature for experimentally determining the MMP.

The advancement of fluid phase behavior prediction and characterization capabilities by a variety of sophisticated, complex, and rigorous equation of states (EOSs) have made it possible MMP to be calculated either analytically or numerically. Both the numerical and analytical methods rely on well-tuned EOS models for determining the MMP and can provide the desired information on the MMP in a relatively faster manner. It is noted here that the term "well-tuned EOS models" refers to the EOS models in which some of the uncertain input parameters have been adjusted for minimizing the differences between the predicted and the measured PVT data. A detailed discussion on the topic of "tuning of EOS models" can be found in published studies like Pederson et al. (1985), Christensen (1999), and Ali and El-Banbi (2015).

Irrespective of the approach (experimental, empirical, analytical, or computational/numerical) taken by an individual operating company for obtaining a reliable information on the MMP for a given CO_2-reservoir oil system, the technical and economic factors to determine the MMP, as discussed in previous sections, justify the necessity to determine the MMP. Now, the question arises as to which approach is best suited to meet the needs of a given project. For answering this question, a basic understanding of the available experimental and non-experimental approaches and their limitations and drawbacks is needed. Also, an appreciation for the underlying physical and thermodynamic principle(s) behind a given method can help in picking an approach that will be able to meet the project's needs. These objectives form the basis of the remaining chapters.

References

Adel IA, Tovar FD, Schechter DS (2016) Fast-slim tube: a reliable and rapid technique for the laboratory determination of MMP in CO_2—light crude oil systems. Soc Pet Eng. https://doi.org/10.2118/179673-MS

Ahmadi K, Johns RT (2011) Multiple-mixing-cell method for MMP calculations. Soc Pet Eng. https://doi.org/10.2118/116823-PA

Ahmadi K, Johns RT, Mogensen K et al (2011) Limitations of current method-of-characteristics (MOC) methods using shock-jump approximations to predict MMPs for complex gas/oil displacements. Soc Pet Eng. https://doi.org/10.2118/129709-PA

Ali MT, El-Banbi AH (2015) EOS tuning—comparison between several valid approaches and new recommendations. Soc Pet Eng. https://doi.org/10.2118/175877-MS

Alomair OA, Garrouch AA (2016) A general regression neural network model offers reliable prediction of CO_2 minimum miscibility pressure. J Petrol Explor Prod Technol 6:351. https://doi.org/10.1007/s13202-015-0196-4

Alomair O, Iqbal M (2014) CO_2 minimum miscible pressure (MMP) estimation using multiple linear regression (MLR) technique. Soc Pet Eng. https://doi.org/10.2118/172184-MS

Alomair O, Malallah A, Elsharkawy A et al (2015) Predicting CO_2 minimum miscibility pressure (MMP) using alternating conditional expectation (ACE) algorithm. Oil Gas Sci Technol Rev. IFP Energies Nouvelles 70 (6):967–982

Alston RB, Kokolis GP, James CF (1985) CO_2 minimum miscibility pressure: a correlation for impure CO_2 streams and live oil systems. SPE J 25(2):268–274

Ashrafizadeh SN, Ghasrodashti AA (2011) An investigation on the applicability of Parachor model for the prediction of MMP using five equations of state. Chem Eng Res Des 89(6):690–696

Ayirala SC, Rao DN (2004) Application of a new mechanistic parachor model to predict dynamic gas-oil miscibility in reservoir crude oil-solvent systems. Soc Pet Eng. https://doi.org/10.2118/91920-MS

Ayirala SC, Rao DN (2006) Comparative evaluation of a new MMP determination technique. Soc Pet Eng. https://doi.org/10.2118/99606-MS

Bui LH, Tsau J-S, Willhite GP (2010) Laboratory investigations of CO_2 near miscible application in Arbuckle Reservoir. Soc Pet Eng. https://doi.org/10.2118/129710-MS

Changlin L, Longxin M, Xianghong W et al (2016) Evaluation method for miscible zone of CO_2 flooding. Electro J Geotech Eng 21:1819–1832

Christensen PL (1999) Regression to experimental PVT data. Petroleum Soc Canada. https://doi.org/10.2118/99-13-52

Christiansen RL, Haines HK (1987) Rapid measurement of minimum miscibility pressure with the Rising-Bubble Apparatus. Soc Pet Eng. https://doi.org/10.2118/13114-PA

Clark NJ, Shearin HM, Schultz WP et al (1958) Miscible drive—its theory and application. Soc Pet Eng. https://doi.org/10.2118/1036-G

Chen BL, Huang HD, Zhang Y et al (2013) An improved predicting model for minimum miscibility pressure (MMP) of CO_2 and crude oil. J Oil Gas Tech 35(2):126–130

Czarnota R, Janiga D, Stopa J et al (2017) Determination of minimum miscibility pressure for CO_2 and oil system using acoustically monitored separator. J CO_2 Utiliz 17:32–36

Czarnota R, Janiga D, Stopa J et al (2017a) Minimum miscibility pressure measurement for CO_2 and oil using rapid pressure increase method. J CO_2 Utiliz 21:156–161

Dong M, Huang SS, Srivastava R (2001) A laboratory study on near-miscible CO_2 injection in Steelman reservoir. Petroleum Soc Canada. https://doi.org/10.2118/01-02-05

Emera MK, Sarma HK (2005) Use of genetic algorithm to estimate CO_2–oil minimum miscibility pressure—a key parameter in design of CO_2 miscible flood. J Pet Sci Eng 46(1–2):37–52

Gamadi TD, Sheng JJ, Soliman MY et al (2014) An experimental study of cyclic CO_2 injection to improve shale oil recovery. Soc Pet Eng. https://doi.org/10.2118/169142-MS

Ghomian Y, Pope GA, Sepehrnoori K (2008) Development of a response surface based model for minimum miscibility pressure (MMP) correlation of CO_2 flooding. Soc Pet Eng. https://doi.org/10.2118/116719-MS

Glaso O (1985) Generalized minimum miscibility pressure correlation (includes associated papers 15845 and 16287). Soc Pet Eng. https://doi.org/10.2118/12893-PA

Gozalpour F, Ren SR, Tohidi B (2005) CO_2 EOR and storage in oil reservoirs. Oil Gas Sci Tech Rev. IFP 60(3):537–546

Hagen S, Kossack CA (1986) Determination of minimum miscibility pressure using a high-pressure visual sapphire cell. Soc Pet Eng. https://doi.org/10.2118/14927-MS

Harmon RA, Grigg RB (1988) Vapor-density measurement for estimating minimum miscibility pressure (includes associated papers 19118 and 19500). Soc Pet Eng. https://doi.org/10.2118/15403-PA

Holm LW, Josendal VA (1974) Mechanisms of oil displacement by carbon dioxide. Soc Pet Eng. https://doi.org/10.2118/4736-PA

Hawthorne SB, Miller DJ, Jin L (2016) Rapid and simple capillary-rise/vanishing interfacial tension method to determine crude oil minimum miscibility pressure: pure and mixed CO_2, methane, and ethane. Energy Fuels 30(8):6365–6372

Jaubert JN, Arras L, Neau E et al (1998a) Properly defining the classical vaporizing and condensing mechanisms when a gas is injected into a crude oil. Ind Eng Chem Res 37:4860

Jaubert J-N, Avaullee L, Pierre C (2002) Is it still necessary to measure the minimum miscibility pressure? Ind Eng Chem Res 41(2):303–310

Jaubert J-N, Wolff L, Neau E et al (1998b) A very simple multiple mixing cell calculation to compute the minimum miscibility pressure whatever the displacement mechanism. Ind Eng Chem Res 37(12):4854–4859

Jessen K, Michelsen ML, Stenby EH (1998) Global approach for calculation of minimum miscibility pressure. Fluid Phase Equilib 153:251

Jin L, Hawthorne S, Sorensen J et al (2017) Advancing CO_2 enhanced oil recovery and storage in unconventional oil play—Experimental studies on Bakken shales. App Energy 208:171–183

Johns RT, Dindoruk B, Orr FM (1993) Analytical theory of combined condensing/vaporizing gas drives. Soc Pet Eng. https://doi.org/10.2118/24112-PA

Johns RT, Orr FM (1996) Miscible gas displacement of multicomponent oils. Soc Pet Eng. https://doi.org/10.2118/30798-PA

Johns RT, Ahmadi K, Dengen Z et al (2010) A practical method for minimum-miscibility-pressure estimation of contaminated CO_2 mixtures. Soc Pet Eng. https://doi.org/10.2118/124906-PA

Johnson JP, Pollin JS (1981) Measurement and correlation of CO_2 miscibility pressures. Soc Pet Eng. https://doi.org/10.2118/9790-ms

Ju B, Wu Y-S, Qin Q (2012) Computer modeling of the displacement behavior of carbon dioxide in undersaturated oil reservoirs. Oil Gas Sci Tech Rev. IFP Energies Nouvelles 70(6):951–965

Karkevandi-Talkhooncheh A, Rostami A, Hemmati-Sarapardeh A et al (2018) Modeling minimum miscibility pressure during pure and impure CO_2 flooding using hybrid of radial basis function neural network and evolutionary techniques. Fuel 220:270–282

Kechut NI, Zain ZM, Ahmad N et al (1999) New experimental approaches in minimum miscibility pressure (MMP) determination. Soc Pet Eng. https://doi.org/10.2118/57286-MS

Khazam M, Arebi T, Mahmoudi T et al (2016) A new simple CO_2 minimum miscibility pressure correlation. Oil Gas Res 2:120. https://doi.org/10.4172/2472-0518.1000120

Lee JI (1979) Effectiveness of carbon dioxide displacement under miscible and immiscible conditions. Report RR-40. Petroleum Recovery Inst., Calgary

Li H, Qin Q, Yang D (2012) An improved CO_2-oil minimum miscibility pressure correlation for live and dead crude oils. Ind Eng Chem Res 51(8):3516–3523

Lorsong J (2013) CO_2 EOR and storage. http://www.ieaghg.org/docs/General_Docs/Summer_School_2013/CO2_EOR__Lorsong_July_13SEC.pdf. Accessed 21 Jan 2018

Mansour EM, Al-Sabagh AM, Desouky SM et al (2017) A new estimating method of minimum miscibility pressure as a key parameter in designing CO_2 gas injection process. Egyptian J Pet. https://doi.org/10.1016/j.ejpe.2017.12.002 (in press)

McCoy ST, Rubin ES (2009) The effect of high oil prices on EOR project economics. Energy Procedia 1(1):4143–4150

Merchant D (2017) Enhanced oil recovery—the history of CO_2 conventional wag injection techniques developed from lab in the 1950's to 2017. Carbon Manag Technol Conf. https://doi.org/10.7122/502866-MS

Metcalfe R (1982) Effects of impurities on minimum miscibility pressures and minimum enrichment levels for CO_2 and rich-gas displacements. Soc Pet Eng. https://doi.org/10.2118/9230-PA

Meyer JP (2007) Summary of carbon dioxide enhanced oil recovery (CO_2EOR) injection well technology. American Petroleum Institute. http://www.api.org/~/media/files/ehs/climate-change/summary-carbon-dioxide-enhanced-oil-recovery-well-tech.pdf. Accessed 22 Jan 2017

Mogensen K, Hood P, Lindeloff N et al (2009) Minimum miscibility pressure investigations for a gas-injection EOR project in Al Shaheen Field. Soc Pet Eng, Offshore Qatar. https://doi.org/10.2118/124109-MS

Mungan N (1981) Carbon dioxide flooding-fundamentals. Petroleum Soc Canada. https://doi.org/10.2118/81-01-03

Nguyen P, Mohaddes D, Riordon J et al (2015) Fast fluorescence-based microfluidic method for measuring minimum miscibility pressure of CO_2 in crude oils. Anal Chem 87(6):3160–3164

Nobakht M, Moghadam S, Gu Y (2008) Determination of CO_2 minimum miscibility pressure from measured and predicted equilibrium interfacial tensions. Ind Eng Chem Res 47:8918–8925

Nouar A, Flock DL (1988) Prediction of the minimum miscibility pressure of a vaporizing gas drive. Soc Pet Eng. https://doi.org/10.2118/15075-PA

Olden P, Mackay E, Pickup G (2015) Techno-economic reservoir modelling final report. Institute of Petroleum Engineering, School of Energy, Geoscience, Infrastructure & Society, Heriot-Watt University. https://www.sccs.org.uk/images/expertise/misc/SCCS-CO2-EOR-JIP-Techno-Economic-Reservoir-Modelling.pdf. Accessed 21 Jan 2018

Orr FM, Jensen CM (1984) Interpretation of pressure-composition phase diagrams for CO_2/crude-oil systems. Soc Pet Eng. https://doi.org/10.2118/11125-PA

Pedersen KS, Thomassen P, Fredenslund A (1985) On the dangers of "tuning" equation of state parameters. Soc Pet Eng. SPE-14487-MS

Qi R, LaForce TC, Blunt MJ (2008) Design of carbon dioxide storage in oil fields. Soc Pet Eng. https://doi.org/10.2118/115663-MS

Qin J, Han H, Liu X (2015) Application and enlightenment of carbon dioxide flooding in the United States of America. Petrol Explor Develop 42(2):232–240

Rezaei M, Eftekhari M, Schaffie M et al (2013) A CO_2-oil minimum miscibility pressure model based on multi-gene genetic programming. Energy Explor Exploit 31(4):607–622

Saini D (2015) CO_2-Prophet model-based evaluation of CO_2-EOR and storage potential in mature oil reservoirs. J Pet Sci Eng 134:79–86

Saini D (2016a) An investigation of the robustness of physical and numerical vanishing interfacial tension experimentation in determining CO_2 + crude oil minimum miscibility pressure. J Petro Eng 2016:13, https://doi.org/10.1155/2016/8150752

Saini D (2017) Engineering aspects of geologic CO_2 storage: synergy between enhanced oil recovery and storage. Springer International Publishing AG, Switzerland

Saini D, Gorecki CD, Knudsen DJ et al (2013) A simulation study of simultaneous acid gas EOR and CO_2 storage at Apache's Zama F Pool. Energy Procedia 37:3891–3900

Saner WB, Patton JT (1986) CO_2 recovery of heavy oil: Wilmington Field test. Soc Pet Eng. https://doi.org/10.2118/12082-PA

Sebastian HM, Wenger RS, Renner TA (1985) Correlation of minimum miscibility pressure for impure CO_2 streams. Soc Pet Eng. https://doi.org/10.2118/12648-PA

Shokir EM (2007) CO_2-oil minimum miscibility pressure model for impure and pure CO_2 streams. J Pet Sci Eng 58(1–2):173–185

Shokrollahi A, Arabloo M, Gharagheizi F et al (2013) Intelligent model for prediction of CO_2—reservoir oil minimum miscibility pressure. Fuel 112:375–384

Shtepani E, Thomas FB, Bennion DB (2006) New approach in gas injection miscible processes modelling in compositional simulation. Petroleum Soc Canada. https://doi.org/10.2118/06-08-01

Srivastava RK, Huang SS (1998) New interpretation technique for determining minimum miscibility pressure by rising bubble apparatus for enriched-gas drives. Soc Pet Eng. https://doi.org/10.2118/39566-MS

Stalkup FI (1978) Carbon dioxide miscible flooding: past, present, and outlook for the future. Soc Pet Eng. https://doi.org/10.2118/7042-PA

Thomas FB, Holowach N, Zhou X et al (1994a) Miscible or near-miscible gas injection, which is better? Soc Pet Eng. https://doi.org/10.2118/27811-MS

Teklu TW, Alharthy N, Kazemi H (2014) Phase behavior and minimum miscibility pressure in nanopores. Soc Pet Eng. https://doi.org/10.2118/168865-PA

Wang S, Ma M, Chen S (2016) Application of PC-SAFT equation of state for CO_2 minimum miscibility pressure prediction in nanopores. Soc Pet Eng. https://doi.org/10.2118/179535-MS

Yang S, Lian L, Yang Y et al (2016) Molecular dynamics simulation of miscible process in co2 and crude oil system. Soc Pet Eng. https://doi.org/10.2118/182907-MS

Zhang K, Jia N, Zeng F et al (2017) A new diminishing interface method for determining the minimum miscibility pressures of light oil–CO_2 systems in bulk phase and nanopores. Energy Fuels 31(11):12021–12034

Valluri MK, Mishra S, Schuetter J (2017) An improved correlation to estimate the minimum miscibility pressure of CO_2 in crude oils for carbon capture, utilization, and storage projects. J Pet Sci Eng 158:408–415

Whorton LP, Brownscombe ER, Dyes AB (1952) Method for producing oil by means of carbon dioxide. US Patent 623,596, 30 Dec 1952

Yellig WF, Metcalfe RS (1980) Determination and prediction of CO_2 minimum miscibility pressures (includes associated paper 8876). Soc Pet Eng. https://doi.org/10.2118/7477-PA

Yin M (2015) CO_2 miscible flooding application and screening criteria. Missouri University of Science and Technology, Thesis

Yuan H, Johns RT (2005) Simplified method for calculation of minimum miscibility pressure or enrichment. Soc Pet Eng. https://doi.org/10.2118/77381-PA

Yuan H, Chopra AK, Marwah V (2008) Fluid characterization for miscible gas floods. Soc Pet Eng. https://doi.org/10.2118/114913-MS

Yuan H, Johns RT, Egwuenu AM et al (2005) Improved MMP correlation for CO_2 floods using analytical theory. Soc Pet Eng. https://doi.org/10.2118/89359-PA

Zhang H, Hou D, Li K (2015) An improved CO_2-crude oil minimum miscibility pressure correlation. J Chem Article ID 175940. https://doi.org/10.1155/2015/175940

Zhang N, Wei M, Bai B (2018) Statistical and analytical review of worldwide CO_2 immiscible field applications. Fuel 220:89–100

Zhao G, Adidharma H, Towler BF et al (2006a) Minimum miscibility pressure prediction using statistical associating fluid theory: two- and three-phase systems. Soc Pet Eng. https://doi.org/10.2118/102501-MS

Zhao GB, Adidharma H, Towler B et al (2006b) Using a multiple-mixing-cell model to study minimum miscibility pressure controlled by thermodynamic equilibrium tie lines. Ind Eng Chem Res 45:7913–7923

Chapter 3
Experimental Approaches

Abstract The three most commonly used experimental approaches currently used by the petroleum industry to characterize and determine the CO_2-reservoir oil miscibility in terms of the MMP are presented and discussed in detail. Their advantages, drawbacks, and limitations as reported in the published literatures are analyzed for improving our current understanding of these experimental approaches.

3.1 Introduction

In the early days, when the petroleum industry had started to recognize the role of miscible drive technology in recovering more oil from reservoirs, Clark et al. (1958) presented a basic framework for determining the general applicability of various types of miscible drive techniques. The main premise of the framework was the recognition of the most suitable type of miscible drive technique to consider, if a given reservoir differs from what is considered ideal. Over time, the petroleum industry became increasingly focused on CO_2 injection-based miscible drive techniques, initially for EOR and later for both EOR and storage while relying on natural as well as anthropogenic sources for a CO_2 supply. Hence, the industry's attention turned to the determination and characterization of CO_2-reservoir oil miscibility in terms of reliable estimates or measurements of the MMP for a given CO_2-reservoir oil system using different (i.e., experimental, analytical, empirical, and computational) approaches.

Obviously, there may be a variety of reasons which would prompt an operating company to take a given approach or a combination of different approaches for determining the MMP for its CO_2-EOR or simultaneous CO_2-EOR and storage project. However, from a business point of view, the cost of the measurements/calculations and time taken for completing the measurements/calculations while ensuring the accuracy and reliability of results obtained from a given approach can be considered as deciding factors, when it comes to select a method for determining the MMP.

It is worth mentioning here that different approaches, which are used to characterize and determine CO_2-reservoir oil miscibility in terms of MMP, all attempt to

D. Saini, *CO₂-Reservoir Oil Miscibility*, SpringerBriefs in Petroleum Geoscience & Engineering, https://doi.org/10.1007/978-3-319-95546-9_3

describe, replicate, and quantify the nature, magnitude, and influence of underlying fluid/fluid (CO_2/reservoir oil) interactions occurring at reservoir conditions of elevated pressure and temperature that is manifested by the MMP even though they occur inside of the pores of the reservoir rock, where another fluid phase (water) is also present.

The simultaneous flow of all three fluid phases through tortuous porous and permeable pathways present in reservoir rock and the wetting preference of the rock surface to oil or water phase (i.e., solid/fluid interactions) also play a role in shaping the interactions between CO_2 and reservoir oil. However, the basic definition of the MMP, which describes the MMP as the pressure at which the local displacement efficiency (i.e., displacement of residual oil by injected CO_2) approaches 100%, takes out the effect of such solid/fluid interactions, while characterizing fluid/fluid interactions taking place between injected CO_2 and reservoir oil. Other possible impacts of solid/fluid interactions on the development of miscibility condition between injected CO_2 and reservoir oil are taken out of consideration by imposing the condition of one-dimensional (1D) flow and no dispersive mixing, while assessing the displacement efficiency, which is exactly 100% at the MMP. According to Wang and Orr (1997), because at pressure near MMP the phases that form in the transition zone between injected fluid and original oil in place have very low interfacial tension, and the displacement is close to piston-like even when small amounts of dispersion or nonuniform flow due to permeability heterogeneity, viscous fingering, or gravity segregation are present, and the conditions of 1D flow and no dispersive mixing are not strictly satisfied.

After taking all the above-mentioned considerations into account, the displacement efficiency essentially becomes a function of fluid/fluid interactions (thermodynamic phase behavior) of the system. Hence, the phenomenon of CO_2-reservoir oil miscibility can essentially be characterized and determined by studying the fluid/fluid interactions that take place between injected CO_2 and reservoir oil at prevailing reservoir conditions of elevated pressure and temperature. Both experimental and non-experimental approaches can be used for studying those interactions.

To study the phenomenon of CO_2-reservoir oil miscibility experimentally, one must either rely on carefully designed displacement experiments for inferring the MMP via achievement of near 100% local displacement efficiency or perform some sort of visual experiment that could provide direct evidence of disappearance of interface between the two phases (CO_2 and reservoir oil). Both—i.e., achievement of near 100% displacement efficiency or disappearance of interface—can provide necessary and sufficient experimental evidence of the achievement of complete miscibility for a given CO_2-reservoir oil system at the reservoir temperature and the pressure, which is referred to as the MMP.

The commonly used experimental methods, which rely on the approach of achieving near 100% displacement efficiency include laboratory core flooding experiments, slim-tube test, and the micro-slim tube test. The frequently used experimental methods, which rely on the experimental approach of directly visualizing the disappearance of the interface for inferring the MMP, include rising bubble apparatus (RBA) and the vanishing interfacial tension (VIT) technique. Both the RBA and VIT techniques, which are the rapid MMP measurement techniques, are frequently used by the industry as an alternative to time-consuming and costly slim-tube tests.

Other experimental methods such as high-pressure visual sapphire cell, single bubble injection technique, vapor liquid equilibrium-interfacial tension test, and PVT multi-contact experiments have also been reported in the published literature for determining and characterizing the CO_2-reservoir oil miscibility, and thus inferring the MMP. However, they are not as popular as the slim-tube test or the RBA and VIT techniques.

Nevertheless, when it comes to experimentally determining and characterizing CO_2-reservoir oil miscibility, the petroleum industry's standard approach is to perform slim-tube tests. The RBA technique, which is almost as old as the slim-tube method, is another well-accepted experimental method for determining and characterizing CO_2-reservoir oil miscibility. The VIT technique has also gained momentum and emerged as a robust experimental technique for determining and characterizing CO_2-reservoir oil miscibility. These three experimental methods are next discussed in detail.

3.2 The Slim-Tube Method

When Whorton et al. (1952) received the first U.S. patent for a method for producing oil by means of CO_2, they relied on the data collected from laboratory displacement tests for deducing the relationship between oil recovery and injection pressure. Subsequently, various researchers (e.g., Rathmell et al. 1971) investigated miscible displacement by CO_2 in the laboratory by conducting displacement of reconstituted reservoir fluids in outcrop cores of varying lengths using CO_2 at various pressures. On the other hand, while conducting laboratory displacement tests to determine the minimum amount of hydrocarbon intermediates to add to a lean gas to achieve miscibility with a given reservoir, Yarborough and Smith (1970) suggested the use of a sufficiently long unconsolidated sand pack (i.e., a sand-packed coil or slime-tube test apparatus) for performing laboratory displacement tests. Their recommendations were based on the rationale that if any porous medium is long, solid/fluid interactions (i.e., relative permeability characteristics) of the porous medium, which do affect component concentration gradients, will not prevent the development of miscibility.

Yellig and Metcalfe (1980) were among early researchers who used the sand-packed coil or slim-tube test apparatus as described by Yarborough and Smith (1970) for determining the CO_2 MMP for reservoir oil. Since then, the slim-tube method has become the industry-accepted approach for experimentally determining the MMP. Typical slim-tube equipment (Figs. 3.1 [after Lim et al. 2008] and 3.2 [after Chen et al. 2017]) are coiled stainless-steel tubes of varying length packed with commercial sand. Other experimental components include transfer vessels, back-pressure regulators, and liquid and gas meters. The sand-packed slim-tube along with other required experimental components is housed in a constant temperature enclosure.

Fig. 3.1 A typical 18 m slim-tube experimental setup, Reprinted by permission: Soc Pet Eng, Design and initial results of EOR and flow assurance lab tests for K2 Field development in the deepwater Gulf of Mexico. Offshore Technology Conference. https://doi.org/10.4043/19624-ms, Lim FH, Munoz E, Joshi NB © 2008

The typical diameter of these coiled-tubes is 0.635 cm [0.25 in], lengths vary from 12 to 36.5 m [40–120 ft] and the porosity and permeability of the packing sand range from 35 to 40% and 2.9×10^{-12} m^2 [3 darcy] to 7.9×10^{-12} m^2 [8 darcy], respectively (Lim et al. 2008).

In the slim-tube method, the cumulative oil recovery factors for the cumulative pore volumes injected, which represent the extent of miscibility between injected CO_2 and reservoir oil, is obtained by performing a series of displacement efficiency tests (i.e., displacement of oil by injected CO_2) at desired test pressures (i.e., operating reservoir pressure [CO_2 injection pressure]) and temperature (i.e., reservoir temperature). A plot (Fig. 3.3) between the collected data (i.e., cumulative recovery factor versus cumulative pore volumes injected) is used for determining the MMP, which is generally the intersection of the two oil recovery trend lines (high slope and low slope) on cumulative oil recovery versus injection pressure plot (Fig. 3.3, after Vulin et al. 2018). A cumulative recovery factor approaching 90% or greater at 1.2 cumulative pore volumes injection is considered a manifestation of the development of miscibility.

As emphasized by Yellig and Metcalfe (1980), a sand-packed coil does not simulate the reservoir rock, instead, it provides a medium for mixing the injected CO_2

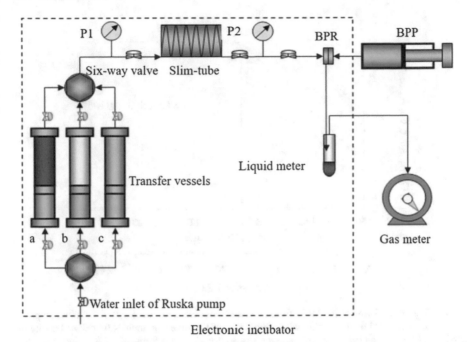

(a) Petroleum ether transfer vessel; (b) Live oil transfer vessel; (c) Gas transfer vessel

Fig. 3.2 Schematic diagram of a typical slim-tube displacement test setup, Reprinted by permission: Taylor & Francis, Decreasing in pressure interval of near-miscible flooding by adding intermediate hydrocarbon components, Geosystem Engineering, Chen H, Tang H, Zhang X et al. © 2017

and reservoir oil in a flowing, multiple-contact process, which is essential for modeling phase behavior or fluid/fluid interactions occurring between CO_2 and oil at given operating (injection) pressure. According to Orr et al. (1982), slim-tube experiments (i.e., displacement tests) do not provide any information about the efficiency of CO_2 floods, even in a laboratory core flooding experiment. Also, it can't be used for deducing information such as CO_2 slug sizes for field applications; instead, it is a simple test procedure that gives immediate information about potential operating reservoir pressure.

As can be seen in Fig. 3.3, a minimum of four to six displacement tests (i.e., two to three tests at operating reservoir pressure below or near the MMP [i.e., immiscible displacement], and two to three tests at operating reservoir pressures near or above the MMP [i.e., miscible displacement]) are needed. The slim-tube method offers a simple experimental test procedure for studying the effect of both the phase behavior and the injection pressure on local displacement efficiency. Hence, it is a critical test that can be performed early in the evaluation of a field prospect for CO_2 flooding (Orr et al. 1982). However, despite its almost universal acceptance, some practitioners have raised concerns regarding experimental ambiguities such

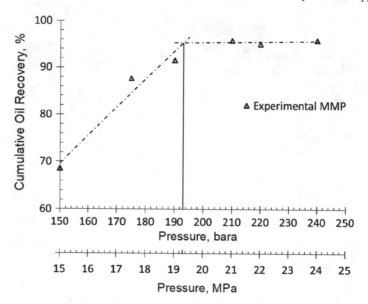

Fig. 3.3 Determination of the MMP from collected cumulative oil recovery versus injection pressure data, Reprinted by permission within the scope of Creative Commons 4.0 license: Faculty of Mining, Geology and Petroleum Engineering University of Zagreb, Slim-tube simulation model for CO_2 injection EOR, Rudarsko-geološko-naftni zbornik (Mining-Geological-Petroleum Engineering Bulletin). https://doi.org/10.17794/rgn.2018.2.4, Vulin D, Gaćina M, Biličić V © 2018

as low cumulative oil recovery at 1.2 pore volumes injected due to channeling or bypassing in a porous medium instead of immiscible displacement and variation in the MMP values recorded by different slim tubes with the same fluid pairs due to poor sand-packing practices (Mogensen et al. 2009).

Various published studies (e.g., Thomas et al. 1994b; Elsharkawy et al. 1996; Dong et al. 2000; Mogensen et al. 2009; Ayirala and Rao 2011; Ekundayo and Ghedan 2013; Voon and Awang 2015; Ahmad et al. 2016) have also pointed out other drawbacks of the use of the slim tube-based experimental approach for determining the MMP. These drawbacks include its time-consuming and costly nature and the lack of standard design and procedure. Due to these drawbacks, the slim-tube method is often labeled a non-unique, uncertain, expensive, and time-consuming method.

According to Ekundayo and Ghedan (2013), the experimental set-up, characteristics of the slim-tube coil (e.g., length, diameter), and experimental procedures (e.g., choice of injection rates) are often left to the discretion of the experimentalist, which may lead to very uncertain and non-unique MMP values. They further point out the lack of measurement repeatability (i.e., two different MMP values for two sets of tests performed using the same fluid samples under the same experimental conditions) in the slim-tube tests. For overcoming such experimental ambiguities, Ekundayo and Ghedan (2013) suggested the use of the longest coil with a moderately large diameter and the use of lowest injection rate possible.

As noted by Ayirala and Rao (2011), one miscibility measurement (i.e., cumulative recovery factor at 1.2 injected pore volumes at a given injection pressure) usually takes several weeks (normally four to five weeks), thus making the slim-tube method quite expensive. Another aspect of the slim-tube method making it expensive is the consumption of a significant amount of representative reservoir oil sample. Both the preparation of a recombined sample of reservoir oil (i.e., recombination of wellhead oil and produced gas samples for achieving reservoir condition gas/oil ratio) in large amounts and/or the collection of live reservoir oil in sufficient quantity via downhole sampling for representing the composition of actual reservoir oil in the experiments add to the cost of the slim-tube method.

Also, the MMP determined from the slim-tube method is an apparent MMP (Yuan et al. 2008) because slim-tube displacements are influenced by physical dispersion, which mostly occurs due to convective type physical mixing of oil and gas phase. According to Johns et al. (2002), oil and gas mixing by physical dispersion and other mechanisms is likely much greater in the reservoir compared to the one observed in laboratory cores. In the case of real reservoirs, various factors such as permeability heterogeneity, viscous fingering, or gravity segregation may augment the effect of physical dispersion or non-uniform flow on the local displacement efficiency, which is typically >90% in slim-tube displacement tests performed at pressures close to the MMP of the CO_2-reservoir oil system investigated. Hence, the scaleup of the physical dispersion in real reservoirs may also be an issue.

Nevertheless, the petroleum industry still relies on the slim-tube method for comparing other experimental and non-experimental methods for their suitability (i.e., robustness, cost-effectiveness, and rapidness) in experimental determination and characterization of CO_2-reservoir oil miscibility in terms of the MMP. Another experimental approach—namely, the rising bubble approach, which is touted as a reliable, fast, and less-expensive experimental alternative to the slim-tube method——is discussed next.

3.3   The Rising-Bubble Method

As documented by Elsharkawy et al. (1996), the rising-bubble method uses a rising-bubble apparatus (RBA) which was developed in the early 1980s by Christiansen and Kim (1986). The RBA apparatus, which is shown schematically in Fig. 3.4 [after Adekunle and Hoffman 2016], consists mainly of a high-pressure sight gauge with a flat glass tube mounted in the center, a bubble injection, three fluid storage/transfer vessels, and a video recording system (Dong et al. 2000).

In the rising-bubble method, MMP is inferred from the pressure dependence of the behavior of the rising bubbles of injected gas through the reservoir oil-filled flat glass tube (Christiansen and Haines 1987). According to Christiansen and Haines (1987), the mass transfer process that occurs as the gas bubble rises through the oil in the glass tube (Fig. 3.5 [after Adekunle and Hoffman 2016]) is similar to the multiple-contact process that arise in slim-tube displacements.

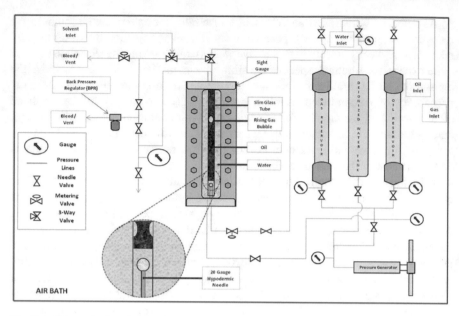

Fig. 3.4 Schematic description of a typical rising-bubble apparatus experimental setup, Reprinted by permission: Elsevier, Experimental and analytical methods to determine minimum miscibility pressure (MMP) for Bakken formation crude oil, J Pet Sci Eng, Adekunle O, Hoffman BT © 2016

As evident from published literature (e.g., Novosad et al. 1990; Thomas et al. 1994b; Elsharkawy et al. 1996; Srivastava et al. 2000; Bon et al. 2005; Sayegh et al. 2007; Torabi and Asghari 2010; Zhang 2016; Li and Luo 2017), the rising-bubble method has been routinely used for experimentally studying the phenomenon of CO_2-reservoir oil miscibility and thus determining the MMP. However, like the slim-tube method, the rising-bubble method has some disadvantages and drawbacks. According to Thomas et al. (1994b), who performed a comparative study of the RBA and slim-tube results, that even though the RBA performed well compared to the slim-tube testing, it provides little quantitative information (e.g., displacement efficiency), and the interpretation of the results is subjective.

Zhou and Orr (1998) presented a theoretical analysis of rising-bubble experiments and concluded that, for vaporizing gas drives, the rising-bubble method can determine the MMP for three-component systems with reasonable accuracy, however, for establishing the reliability of the rising-bubble method in determining and characterizing CO_2-reservoir oil miscibility for multicomponent systems that show mixed-drive (i.e., vaporizing/condensing) behavior, additional work is required.

Recently, Li and Luo (2017) have investigated the use of the RBA method for determining the MMP for a Bakken tight oil and different injection gases. They also performed EOS simulations for determining the mechanism that was responsible for developing miscibility in those CO_2-reservoir oil systems. For a pure CO_2-Bakken oil system, one EOS simulation method (cell-to-cell) predicted that the vaporizing

(a)   (b)   (c)

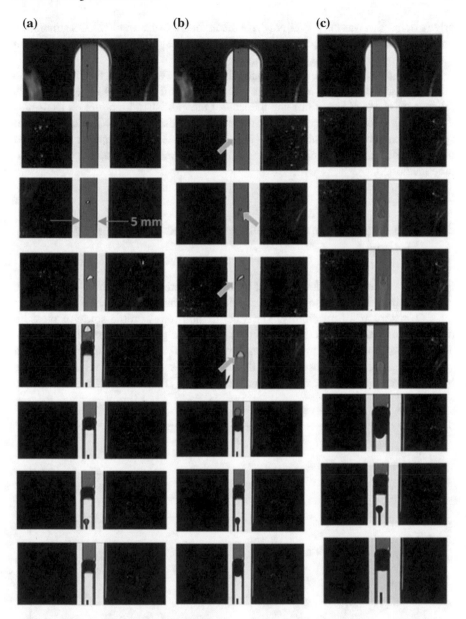

Fig. 3.5 Rising of CO_2 bubble through the dead reservoir oil at different experiment pressure steps (**a** experimental pressure < MMP, **b** experimental pressure @ MMP, **c** experimental pressure > MMP), Reprinted by permission: Elsevier, Experimental and analytical methods to determine minimum miscibility pressure (MMP) for Bakken formation crude oil, J Pet Sci Eng, Adekunle O, Hoffman BT © 2016

drive mechanism was the prevailing drive mechanism, while another EOS simulation method (multiple-mixing-cell) concluded that it was the condensing drive mechanism that led to the development of miscibility.

As evident from the above-mentioned discussion, even though the rising-bubble method is quite rapid and inexpensive and needs small quantities of test fluids, its reliance on visual observations and the lack of the published literature on its use for determining and characterizing CO_2-reservoir oil miscibility for multicomponent systems that show mixed-drive (i.e., vaporizing/condensing) behavior have posed obstacles for making it a universally accepted experimental alternative of the slim-tube method.

3.4   The Experimental Vanishing Interfacial Tension (VIT) Method

Rao (1997) reported the development of a new experimental method—namely the vanishing interfacial tension (VIT) technique for characterizing and determining the CO_2-reservoir oil miscibility. The VIT method relies on the study of pressure dependence of the interfacial tension (IFT) behavior of CO_2-reservoir oil system. The experimentally-measured CO_2-reservoir oil IFT at reservoir temperature and varying pressure steps is used for making an IFT versus pressure versus plot. The MMP is then obtained by extrapolating the observed IFT versus 1/pressure trend to

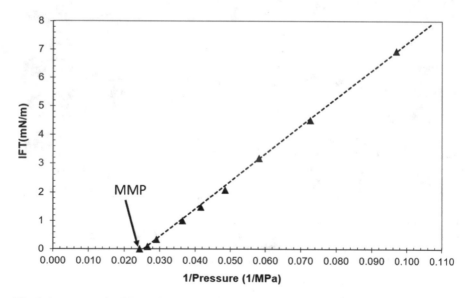

Fig. 3.6 An example of determining the MMP from experimentally-obtained IFT versus 1/Pressure curve by using the VIT method

zero IFT (Fig. 3.6). It is worth mentioning here that the pressure dependence of the IFT behaviors of CO_2-reservoir oil systems can be subdivided in two distinct trends, one in low pressure or high IFT region (i.e., IFT > 1 mN) and another in high pressure or low IFT region (i.e. IFT < 1 mN/m). The use of IFT versus 1/pressure curve helps in linearizing these two distinct trends for obtaining the MMP by extrapolation of this linear trend to a condition of zero IFT.

It is well known that at sufficiently low IFT values (less than 0.1 mN/m and perhaps as low as 0.001 mN/m), displacement efficiencies under immiscible conditions can approach those for miscible floods (Hsu et al. 1985). As emphasized by Nagarajan et al. (1990), the near-critical region where IFT approaches zero is of particular interest in EOR applications, since improved efficiency of oil displacement by CO_2 occurs when the IFT becomes low. According to Ayirala and Rao (2011), in the VIT method, multiple contact interactions between injected gas and reservoir oil are studied in such a manner that achievement of zero IFT condition corresponds to the MMP.

3.4.1 Experimental Data Collection and Interpretation Methodology

A high-pressure high-temperature (HPHT) optical cell system (Fig. 3.7), which facilitates the use of both the pendant drop (or falling drop) method and the capillary rise method for determining the CO_2-reservoir oil IFT via the analysis of captured images (i.e., pendant oil drop shape profiles and the height of capillary rise of oil-rich phase) resulting from the multiple contact interactions between equilibrated oil-rich and CO_2-rich phases, is used for experimentally measuring the CO_2-reservoir oil IFT at reservoir temperature and varying pressure steps. The presence of a light source on one side and a video camera on the opposite side of the HPHT optical cell's glass windows facilitates the capturing and recording of drop shapes and capillary rise heights on a video recording system.

In the VIT experimental procedure, first, CO_2 is charged into the preheated HPHT cell maintained at the reservoir temperature and desired experimental pressure. Next, reservoir oil, which is already at the reservoir temperature and experimental pressure conditions, is introduced into the CO_2-filled HPHT cell to form a pendant drop immersed in CO_2 phase using the metal capillary tube available at the top of the HPHT optical cell. The drop profile images of first few pendant oil drops immersed in CO_2-phase are captured on a video tape for later analysis of the shape profiles using a drop shape analysis method such as the ADSA (axisymmetric drop shape analysis) technique for determining the first-contact IFT at experimental pressure and reservoir temperature. It is worth mentioning here that the measured "first-contact" IFT quantifies the interactions occurring in the reservoir in which freshly injected CO_2 phase encounters the fresh reservoir oil for the first time.

Now, a small amount (5–7 ml) of reservoir oil, sufficient to observe a rise of oil phase in the glass capillary tube already stationed inside the HPHT optical cell, is added in a drop-wise manner for allowing multiple-contact interactions and for obtaining equilibrated oil-rich and CO_2-rich phases while obtaining a capillary rise of oil phase in the glass capillary tube (Fig. 3.8). After an aging period ranging from 1 to 3 h, the height of capillary rise of the equilibrated oil phase is imaged and recorded. Both equilibrated oil-rich and CO_2-rich phases are also sampled for obtaining densities as well as compositions of the equilibrated phases.

Next, fresh CO_2 is introduced for elevating the pressure in the HPHT optical cell to the next experimental pressure step. Again, after fresh reservoir oil in introduced to form few pendant drops, the interactions between newly introduced oil phase and already available oil-rich and CO_2-rich phases are not first-contact interactions anymore. The above-mentioned process signifies the interactions occurring in the reservoir between the freshly injected CO_2 phase and previously contacted reservoir oil (i.e., oil-rich phase) and/or already contacted CO_2 (i.e., CO_2-rich phase) and fresh reservoir oil. The interactions are essentially multiple-contact interactions which are quantified in terms of multiple-contact dynamic IFT.

Hence, an analysis of the captured drop shape profiles and the observed capillary rise heights, now, results in multiple-contact dynamic IFT at experimental pressure and reservoir temperature. The equilibrated oil-rich and CO_2-rich phases are sampled again for obtaining their densities and compositions at experimental pressure and reservoir temperature. The procedure is repeated at several other higher experimental pressures until no more regular shaped-pendant drops of reservoir oil can be formed. The formation of the irregular shaped-pendant oil drop itself is an indication that the CO_2-reservoir oil system is approaching a low IFT region and is about to achieve the condition of miscibility. At this stage, only capillary rise height is measured as pendant drop technique is no more suitable for determining the IFT between oil-rich and CO_2-rich phases.

It is worth mentioning here that, when Rao (1997) published the first study on the use of the VIT method for determining the MMP, he used the pendant drop (or falling-drop) technique for measuring the CO_2-reservoir oil IFT at several pressure steps. The pendant drop (or falling-drop) technique is a commonly used method, in which the shape profiles of the pendant drops of the liquid phase immersed in the gas phase are captured and analyzed. Later, Ayirala (2005), Sequeira (2006), and Saini and Rao (2010) used a combination of both the pendant drop and the capillary rise techniques to study the pressure dependence of the IFT behavior of CO_2-reservoir oil systems and thus characterizing and determining the MMP using the VIT method. They have also provided detailed discussions on the experimental and calculation procedures employed by them while performing the IFT measurements.

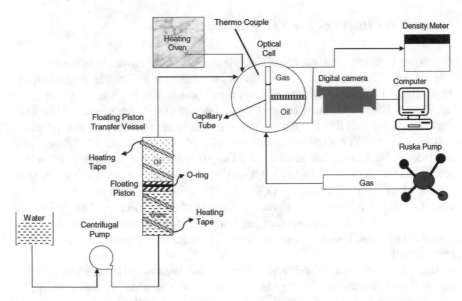

Fig. 3.7 Schematic description of the VIT experimental setup, Reprinted by permission: Soc Pet Eng, Comparative evaluation of a new gas/oil miscibility-determination technique, https://doi.org/10.2118/99606-pa, Ayirala SC, Rao DN © 2011

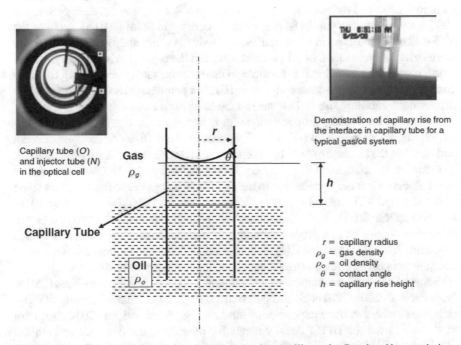

Fig. 3.8 Schematic description of rise of oil phase in glass capillary tube, Reprinted by permission: Soc Pet Eng, Comparative evaluation of a new gas/oil miscibility-determination technique, https://doi.org/10.2118/99606-pa, Ayirala SC, Rao DN © 2011

3.4.2 Robustness of the VIT Method

According to Ayirala and Rao (2011), as injected gas [CO_2] advances in the reservoir, it continues to interact with fresh reservoir oil residing ahead of the injection front. Meanwhile, the fresh CO_2 being injected continuously into the injection well meets the residual reservoir oil that has previously interacted with the injected CO_2. They emphasized the fact that the placement of a certain amount of reservoir oil at the bottom of the HPHT optical cell allows the injected CO_2 to get into equilibrium with reservoir oil for forming an oil-rich and a CO_2-rich phase. Fresh reservoir oil is then brought in contact with equilibrated CO_2-rich phase by forming pendant drops from the top of the HPHT cell, which represents the scenario of dynamic displacement occurring in the reservoir. This process, mimicked in the VIT method, is like the one that occurs in the reservoir, called "multicontact" because the injected CO_2 continually interacts with residual reservoir oil and moves forward to contact fresh reservoir oil.

Recently, Saini (2016a) reviewed various VIT method-based experimental studies reported in the published literature to investigate the robustness of both the physical and numerical VIT experimentation in characterizing and determining the CO_2-reservoir oil miscibility in terms of the MMP. Table 3.1 provides a summary of experimental apparatus, CO_2-reservoir oil system(s) investigated, IFT measurement technique used, and the results obtained in different VIT method-based experimental studies reported in the published literature. According to Saini (2016a), some of the VIT method-based experimental studies used only the pendant drop method whereas others used a combination of the pendant drop and the capillary rise techniques. It is worth mentioning here that it is possible to characterize and quantify both the first-contact and the multiple-contact dynamic IFT, if a combination of both the pendant drop and the capillary rise techniques is used in a single experiment.

The use of the pendant drop technique for measuring CO_2-reservoir oil IFT without placing a small amount of initial oil appears to be a "first-contact" experiment and signifies the absence of "multiple contact" interactions (e.g., Awari-Yusuf 2013; Nobakht et al. 2008; Patil et al. 2008). Hence the measured IFT versus pressure trend observed is more correlatable to the FC-MMP compared to the MMP. In some of the reported VIT method-based experimental studies (e.g., Saini and Rao 2010; Ayirala and Rao 2011), the CO_2-reservoir oil IFT measurements at various pressure steps were conducted in a single experiment compared to other experimental studies (Sequeira et al. 2008) where CO_2-reservoir oil IFT at various pressure steps were measured via performing a separate experiment at each pressure step.

Some of the VIT method-based experimental studies (e.g., Awari-Yusuf 2013; Nobakht et al. 2008; Patil et al. 2008) have used the EOS-calculated density of CO_2-rich phase, whereas, other experimental studies (e.g., Saini and Rao 2010; Sequeira et al. 2008) used the HPHT density meter for measuring the densities of both the equilibrated oil-rich and CO_2-rich phases. Sequeira et al. (2008) also performed the detailed compositional analysis of the collected oil-rich and CO_2-rich phases. Such compositional analysis, if performed in an VIT method-based experimental study,

Table 3.1 Summary of experimental apparatus, CO₂-reservoir oil system(s) investigated, IFT measurement technique used, and key results reported in the published VIT method-based experimental studies (modified by permission from Saini 2016a)

Reported study	Experimental apparatus used	System investigated	IFT measurement technique(s) employed	Method(s) used for measuring gas and oil phase densities	Results obtained
Rao (1997)	High-pressure high-temperature optical cell	Hydrocarbon gas + live crude oil; Hydrocarbon gas + dead crude oil	Pedant drop	Oil and gas: equation of state	MMP, FC-MMP, MMC
Yang and Gu (2004)	See-through windowed high-pressure cell	CO_2 + dead crude oil	Pedant drop	Gas: standard property table oil: atmospheric dead oil density	FC-MMP
Ayirala (2005), Ayirala and Rao (2006)	High-pressure high-temperature optical cell	CO_2 + n-Decane; CO_2 + live decane	Pedant drop and capillary rise methods	Oil and gas: density meter	MMP
Sequeira (2006), Sequeira et al. (2008)	High-pressure high-temperature optical cell	CO_2 + live crude oil	Pedant drop and capillary rise methods	Oil and gas: density meter	MMP, FC-MMP
Nobakht et al. (2008)	See-through windowed high-pressure cell	CO_2 + dead crude oil	Pedant drop	Gas: equation of state oil: atmospheric dead oil density	FC-MMP
Patil et al. (2008)	High pressure optical cell	CO_2, CH_4 or viscosity reducing injectant) + ANS crude oil	Pedant drop	Oil and gas: equation of state	MMP
Wang et al. (2010)	See-through windowed high-pressure cell	CO_2 and three different crude oils	Pedant drop	Gas: equation of state oil: N/A	"so-called" MMP, "so-called" FC-MMP
Saini and Rao (2010)	High-pressure high-temperature optical cell	CO_2 + live crude oil	Pedant drop and capillary rise methods	Oil and gas: density meter	MMP
Awari-Yusuf (2013)	–	CO_2 + dead crude oil	Pedant drop	Gas: equation of state oil: pycnometer	FC-MMP

would be sufficient for deducing information about the prevailing drive mechanism(s) (vaporizing, condensing, or mixed) responsible for attaining the miscibility condition as well.

3.4.3 Theoretical Concerns About the Robustness of the VIT Method

Based on the theoretical treatment of the phenomenon of miscibility, researchers have expressed concerns about the reliability of the VIT method in characterizing and determining the CO_2-reservoir oil miscibility by studying pressure dependence of the IFT behaviors of CO_2-reservoir oil systems. Orr and Jessen (2007) observed the absence of interface and its relevance to the condition of miscibility. They used the term "dissolution" to describe the situation in which a small amount of oil and gas can be added to a mixture without forming two phases. The absence of interface at such conditions (i.e., bubble point pressure and dewpoint pressure) raises a serious

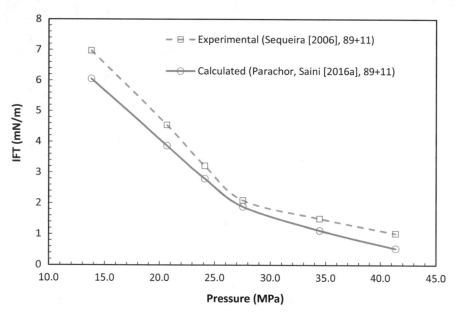

Fig. 3.9 The experimental IFT (Sequeira 2006) and the calculated (Parachor) IFT behaviors (Saini 2016a) at 114.4 °C (238 °F) for constant mol ratio (89 + 11) feed mixture of CO_2-live reservoir oil system, Reproduced by permission within the scope of open access article: OMICS International, Modeling of pressure dependence of interfacial tension behaviors of a complex supercritical CO_2 + live crude oil system using a basic Parachor expression, J Pet Environ Biotechnol 7:277 Saini D © 2016

concern about the suitability of the VIT method (a condition of zero IFT or in other words no interface separating the phases) for determining the miscibility.

Based on their simulated VIT experiments, Jessen and Orr (2008) concluded that the VIT experiments do not, in general, create critical mixtures relevant to the development of multicontact miscibility in gas/oil displacement processes, and the mixture composition used in a VIT experiment has a significant impact on the accuracy of the estimated MMP relative to the experimental slim-tube MMP. They further stated that VIT experiment estimates of the MMP obtained by extrapolating IFTs from low values of the IFT are least accurate for gas/oil systems that have an FC-MMP much higher than the MMP. According to them, additional experimental information would be required to select the optimal cell-mixture composition that would give a reasonably accurate estimate of the MMP by extrapolation of the IFT versus pressure curve to the condition of zero IFT. In other words, if the VIT method is used for determining the MMP for any feed mixture composition which does not correspond to a so-called "optimal feed mixture composition," the VIT method's results (i.e., MMP values) would significantly differ from the MMP values derived from the slim-tube method.

On the other hand, in the light of the above-mentioned theoretical concerns, Sequeira et al. (2008) examined the effect of feed mixture composition pressure dependence of the IFT behavior of a complex CO_2-reservoir oil system. They per-

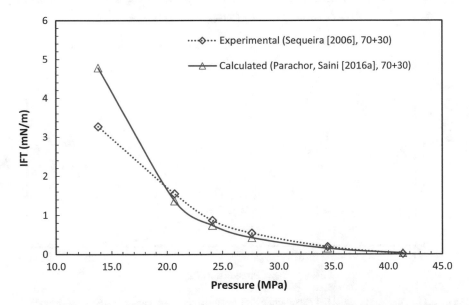

Fig. 3.10 The experimental IFT (Sequeira 2006) and the calculated (Parachor) IFT behaviors (Saini 2016a) at 114.4 °C (238 °F) for constant mol (70 + 30) feed mixture of CO_2- live reservoir oil system, Reproduced by permission within the scope of open access article: OMICS International, Modeling of pressure dependence of interfacial tension behaviors of a complex supercritical CO_2 + live crude oil system using a basic Parachor expression, J Pet Environ Biotechnol 7:277 Saini D © 2016

formed VIT experiments using two constant mole compositions (i.e., 70 + 30, and 90 + 10 mol% of CO_2 and reservoir oil) and two constant volume (i.e., 45 + 55, and 85 + 15 volume% of CO_2 and reservoir oil) feed mixtures at a reservoir temperature of 114.4 °C (238 °F). It is worth mentioning here that the constant volume feed mixtures tested by them are also equivalent to variable mole composition feed mixtures. In the case of the first variable composition feed mixture, CO_2 composition varied from 88 to 95 mol% (i.e., 12–5 mol% reservoir oil). For the second mixture, CO_2 composition varied from 52 to 74 mol% (i.e., 48–26 mol% reservoir oil). Although the pressure dependence of the IFT behaviors varied with feed mixtures' compositions, all of them eventually converge nearly to the same point of zero IFT yielding similar MMPs. In other words, the MMP deduced from the VIT method was not affected by the feed mixtures' composition.

Obviously, theoretical treatment of the VIT method reported in published literature provides an opportunity to test the robustness of the VIT method in determining and characterizing the phenomenon of CO_2-reservoir oil miscibility accurately. The experimental results presented by Sequeira (2006), Sequeira et al. (2008) have already addressed majority of theoretical concerns about the VIT method.

Recently, Ahmad et al. (2016) have also compared the MMP values derived by the VIT method with the MMP values obtained from the slim-tube method for three

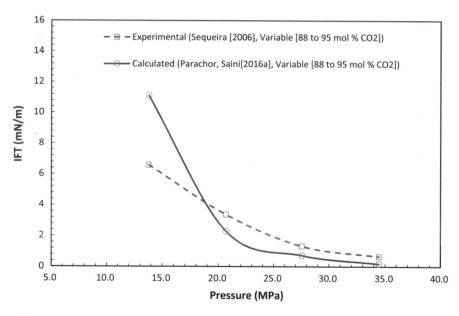

Fig. 3.11 The experimental IFT (Sequeira 2006) and the calculated (Parachor) IFT behaviors (Saini 2016a) at 114.4 °C (238 °F) for variable mol (88–95 mol% CO_2) feed mixture of CO_2-live reservoir oil system, Reproduced by permission within the scope of open access article: OMICS International, Modeling of pressure dependence of interfacial tension behaviors of a complex supercritical CO_2 + live crude oil system using a basic Parachor expression, J Pet Environ Biotechnol 7:277 Saini D © 2016

CO_2-reservoir oil systems. They concluded that since the estimated MMPs by the VIT method are in close match with more reliable slim-tube method's results, the VIT method can be used as a reliable and cheap alternative to the more expensive and time-consuming slim-tube method for accurate MMP determination.

The results of Sequeira et al. (2008) and Ahmad et al. (2016) demonstrate that the VIT method is successful in characterizing the CO_2-reservoir oil miscibility in such a manner that the achievement of zero IFT condition corresponds to the MMP. On the other hand, the apparent dependence of the VIT MMP on feed mixture composition seems to arise from the common limitation of the EOS models in predicting the equilibrium fluid phase densities in the near-critical region. This issue is discussed next.

3.4.4 Pressure Dependence of the IFT Behaviors of CO_2-Reservoir Oil Systems

The Parachor-based basic expressions are often used to model the experimentally observed pressure dependence of the IFT behaviors of complex CO_2-reservoir oil

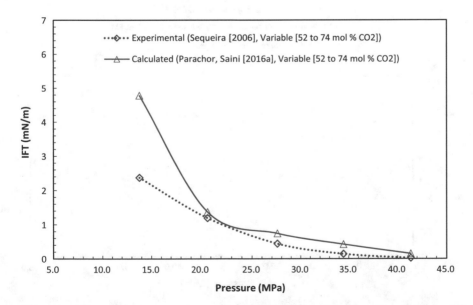

Fig. 3.12 The experimental IFT (Sequeira 2006) and the calculated (Parachor) IFT behaviors (Saini 2016a) at 114.4 °C (238 °F) for variable mol (52–74 mol% CO_2) feed mixture of CO_2-live reservoir oil system, Reproduced by permission within the scope of open access article: OMICS International, Modeling of pressure dependence of interfacial tension behaviors of a complex supercritical CO_2 + live crude oil system using a basic Parachor expression, J Pet Environ Biotechnol 7:277 Saini D © 2016

(dead or live) mixtures at elevated temperatures. However, such modeling requires compositions and densities of the equilibrium liquid and vapor phases and molecular weights of various components present in the system, and Parachors for pure components and heavier hydrocarbon components (C_{7+} fractions). Unfortunately, the experimentally-measured input data needed to perform Parachor expression-based modeling are not readily available in the published literature. In the absence of experimentally-measured data, available phase behavior packages are often used for obtaining such input data as needed for modeling (e.g., Jessen and Orr 2008; Al Riyami and Rao 2015). The use of experimentally-measured input data for performing Parachor-based modeling of the pressure dependence of the IFT behaviors of complex CO_2-reservoir oil systems are almost non-existent in the published literature.

Recently, Saini (2016b) used a Parachor-based basic expression for modeling the pressure dependence of the IFT behavior of a complex CO_2-reservoir oil system for which experimentally-measured input data (compositions and densities of the equilibrium CO_2-rich and oil-rich phases and molecular weights of various components present in the system) and pressure dependence of the IFT behavior curves, and the

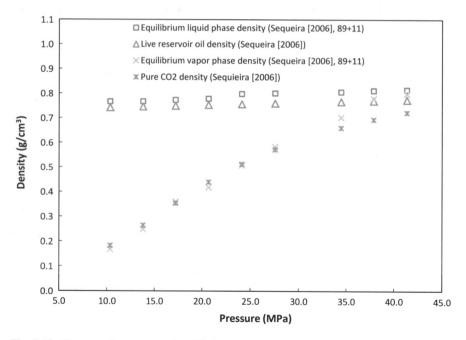

Fig. 3.13 Corresponding measured equilibrium and pure phase densities (Sequeira 2006) to the experimental IFT behavior at 114.4 °C (238 °F) for constant mol ratio (89 + 11) feed mixture of CO_2-live reservoir oil system, Reproduced by permission within the scope of open access article: OMICS International, Modeling of pressure dependence of interfacial tension behaviors of a complex supercritical CO_2 + live crude oil system using a basic Parachor expression, J Pet Environ Biotechnol 7:277 Saini D © 2016

VIT method's derived MMPs were available via the published literature (Sequeira 2006; Sequeira et al. 2008). Saini (2016b) used equilibrium liquid phase (C_{7+} fraction) molecular weights measured at various pressures reported by Sequeira (2006) for determining the C_{7+} fraction Parachor using a correlation presented by Schechter and Guo (1998).

A comparison of the experimentally-measured pressure dependence of the IFT behaviors of different feed mixtures (Sequeira 2006; Sequeira et al. 2008) and the basic Parachor expression-based calculated pressure dependence of the IFT behaviors (Saini 2016a), shown in Figs. 3.9, 3.10, 3.11 and 3.12, suggest that, for all four feed mixtures, the calculated IFT behaviors in high pressure (low IFT i.e. <1 mN/m) regions appeared to follow the experimental IFT behaviors reasonably well. As expected, the deviation was more pronounced in a low-pressure (high IFT i.e. >1 mN/m) region because basic Parachor expressions are mainly used for predicting IFT behaviors in a near-critical region, where IFT between the phases starts to diminish.

Also, the difference between the calculated and the experimental IFT behaviors appears to diminish in low IFT (<1mN/m) regions for all (Figs. 3.9, 3.10, 3.11 and 3.12) but one (Fig. 3.9) feed mixtures. On the other hand, in the low pressure

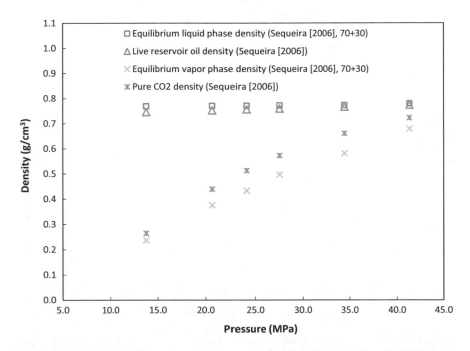

Fig. 3.14 Corresponding measured equilibrium and pure phase densities (Sequeira 2006) to the experimental IFT behavior at 114.4 °C (238 °F) for constant mol (70 + 30) feed mixture of CO_2-live reservoir oil system, Reproduced by permission within the scope of open access article: OMICS International, Modeling of pressure dependence of interfacial tension behaviors of a complex supercritical CO_2 + live crude oil system using a basic Parachor expression, J Pet Environ Biotechnol 7:277 Saini D © 2016

(high IFT [>1 mN/m]) region, even though the densities of the equilibrium phases showed a little deviation from densities of corresponding pure phases (Figs. 3.13, 3.14, 3.15 and 3.16), a significant drop in IFT values can easily be observed. This can be attributed to the intermolecular interactions occurring between the CO_2-rich and oil-rich phases without any significant mass transfer between the equilibrium phases.

As can be seen from Figs. 3.13, 3.14, 3.15 and 3.16, in high-pressure regions, density differences between the phases (CO_2-rich and oil-rich phases; pure reservoir oil and CO_2) reduced rapidly. This indicates that mass transfer between the phases dramatically increased as the system moved closer to near-critical pressure condition. As expected, calculated IFTs started to follow the experimentally-measured IFT values more closely. However, in a high-pressure (IFT < 1mN/m) region, IFT decline rate was quite low compared to a sharp decline rate observed in low-pressure (IFT > 1 mN/m) regions.

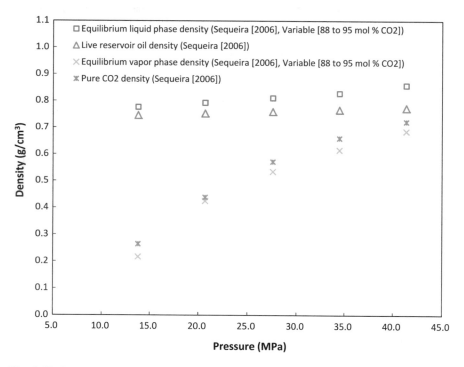

Fig. 3.15 Corresponding measured equilibrium and pure phase densities (Sequeira 2006) to the experimental IFT behavior at 114.4 °C (238 °F) for variable mol (88–95 mol% CO_2) feed mixture of CO_2-live reservoir oil system, Reproduced by permission within the scope of open access article: OMICS International, Modeling of pressure dependence of interfacial tension behaviors of a complex supercritical CO_2 + live crude oil system using a basic Parachor expression, J Pet Environ Biotechnol 7:277 Saini D © 2016

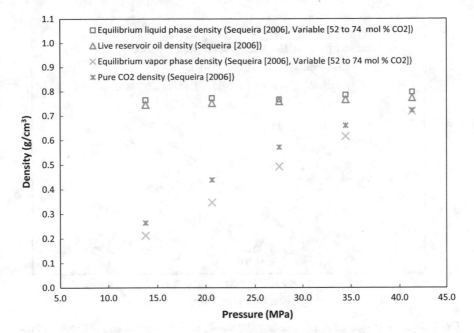

Fig. 3.16 Corresponding measured equilibrium and pure phase densities (Sequeira 2006) to the experimental IFT behavior at 114.4 °C (238 °F) for variable mol (52–74 mol% CO_2) feed mixture of CO_2-live reservoir oil system, Reproduced by permission within the scope of open access article: OMICS International, Modeling of pressure dependence of interfacial tension behaviors of a complex supercritical CO_2 + live crude oil system using a basic Parachor expression, J Pet Environ Biotechnol 7:277 Saini D © 2016

As evident from IFT versus 1/Pressure plots shown in Figs. 3.17, 3.18, 3.19 and 3.20, both the experimental and the calculated curves, irrespective of a large variation in the feed mixture compositions, appeared to converge on a very similar pressure when extrapolated to zero IFT condition. Also, the deviation between the calculated pressure and the experimentally-determined pressure values corresponding to zero IFT condition (i.e., the calculated MMP and the experimentally-determined MMP values using the VIT method) ranged from +5.5% to −8.3% (Table 3.2). These results demonstrate that the experimentally-determined MMP using the VIT method does not depend on the feed mixture composition. In other words, a knowledge of a so-called "optimal feed mixture composition" is not an a priori condition for using the VIT method for determining the MMP.

As evident from above-mentioned studies (i.e., Saini 2016a; Sequeira 2006; Sequeira et al. 2008), in the low IFT region (<1 mN/m), both the calculated and the experimental IFT versus 1/pressure curves can be used for obtaining reasonable estimates of the MMP, if experimentally-measured input data (i.e., compositions and densities of the equilibrium CO_2-rich and oil-rich phases and molecular weights of various components present in the system) is used.

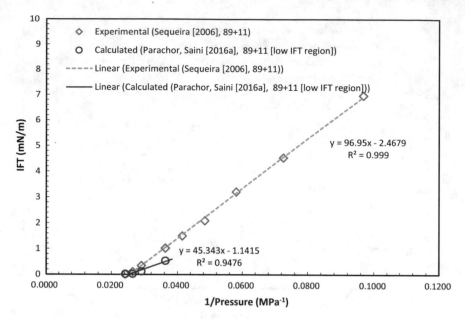

Fig. 3.17 The experimental IFT (Sequeira 2006) and the calculated (Parachor) IFT versus 1/Pressure trends (Saini 2016a) at 114.4 °C (238 °F) for constant mol ratio (89+11) feed mixture of CO_2-live reservoir oil system, Reproduced by permission within the scope of open access article: OMICS International, Modeling of pressure dependence of interfacial tension behaviors of a complex supercritical CO_2 +live crude oil system using a basic Parachor expression, J Pet Environ Biotechnol 7:277 Saini D © 2016

3.4.5 Prediction of Prevailing Drive Mechanism(s)

In the VIT method, reservoir oil and CO_2 are brought in contact with each other in a manner that the molecular interactions taking place between different phases (i.e., pure CO_2 and pure reservoir oil; CO_2-rich phase and pure reservoir oil; pure CO_2 and oil-rich phase; CO_2-rich phase and oil-rich phase) in the reservoir during the displacement of reservoir oil by injected CO_2 are reasonably depicted. On the other hand, for determining the IFT of a CO_2-reservoir oil system at given experimental pressure, densities of both the CO_2-rich and the oil-rich phases when both phases are in equilibrium condition, needs to be known. In the VIT method, densities of both the CO_2-rich and the oil-rich phases are experimentally measured. The measured density data (e.g., Figs. 3.13, 3.14, 3.15 and 3.16) can also be used for deducing useful information about the prevailing drive mechanism(s) responsible for miscibility development.

As can be seen in Figs. 3.13, 3.14, 3.15 and 3.16, densities of both the equilibrated CO_2-rich (vapor) and oil-rich (liquid) phases did not deviate too much from the density values of pure phases (CO_2 and reservoir oil) until the experimental pressure reached close to MMP (or close to critical pressure of the system). At experimental

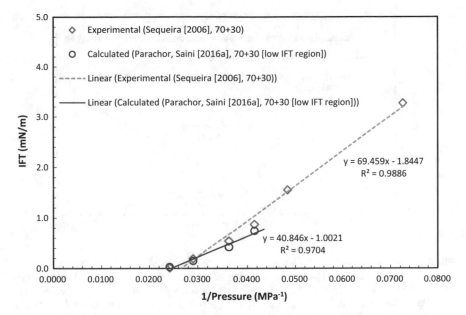

Fig. 3.18 The experimental IFT (Sequeira 2006) and the calculated (Parachor) IFT versus 1/Pressure trends (Saini 2016a) at 114.4 °C (238 °F) for constant mol (70 + 30) feed mixture of CO_2- live reservoir oil system, Reproduced by permission within the scope of open access article: OMICS International, Modeling of pressure dependence of interfacial tension behaviors of a complex supercritical CO_2 + live crude oil system using a basic Parachor expression, J Pet Environ Biotechnol 7:277 Saini D © 2016

pressure steps that were close to the MMP, the density of the CO_2-rich phase showed a large deviation while density of oil-rich phase showed only a gradual but small deviation from their pure phase density values. Such behavior is an indication of a vaporizing drive mechanism because it was extraction (i.e., vaporization) of certain lighter hydrocarbon components from reservoir oil by CO_2 that caused a significant increase in CO_2-rich phase's density due to increased mass transfer in low IFT (< 1 mN/m). Obviously, the oil-rich phase was left with heavier hydrocarbon components which resulted in a gradual but small increase in its density compared to pure reservoir oil phase's density. Further, availability of compositional data of both the equilibrated CO_2-rich and oil-rich phases (e.g., Sequeira 2006; Sequeira et al. 2008) can also be used for reaffirming the prevailing drive mechanism(s) by comparing the extent of transfer of different components (i.e., percent increase or decrease with respect to their initial compositions) between the two phases.

If similar observations are made for deducing information about the prevailing drive mechanism(s) responsible for development of miscibility in a simple CO_2-hydrocabon system—namely CO_2-n-decane system investigated by Nagarajan and Robison (1986)—the equilibrated phase density and IFT behavior data indicates the presence of both vaporizing (i.e., extraction of decane by CO_2 phase resulting in a sharp increase in the density of the CO_2-rich phase compared to the density of a pure

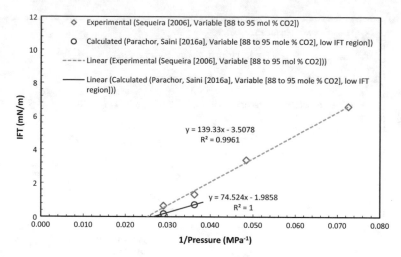

Fig. 3.19 The experimental IFT (Sequeira 2006) and the calculated (Parachor) IFT versus 1/Pressure trends (Saini 2016a) at 114.4 °C (238 °F) for variable mol (88–95 mol% CO_2) feed mixture of CO_2-live reservoir oil system, Reproduced by permission within the scope of open access article: OMICS International, Modeling of pressure dependence of interfacial tension behaviors of a complex supercritical CO_2 + live crude oil system using a basic Parachor expression, J Pet Environ Biotechnol 7:277 Saini D © 2016

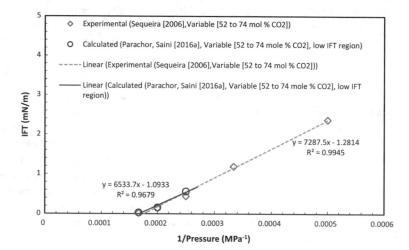

Fig. 3.20 The experimental IFT (Sequeira 2006) and the calculated (Parachor) IFT versus 1/Pressure trends (Saini 2016a) at 114.4 °C (238 °F) for variable mol (52–74 mol% CO_2) feed mixture of CO_2- live reservoir oil system, Reproduced by permission within the scope of open access article: OMICS International, Modeling of pressure dependence of interfacial tension behaviors of a complex supercritical CO_2 + live crude oil system using a basic Parachor expression, J Pet Environ Biotechnol 7:277 Saini D © 2016

Table 3.2 Comparison of the Parachor model-based calculated MMPs (Saini 2016b) with the experimentally-measured VIT MMPs for different CO_2-reservoir oil mixtures investigated by Sequeira (2006) and reported by Sequeira et al. (2008)

System	System type	Deviation between the experimentally-measured VIT MMP (Sequeira 2006; Sequeira et al. 2008 and the Parachor model-based MMP Saini 2016b) (%)
89 + 11	Constant mol ratio	−1.0
70 + 30	Constant mol ratio	−8.3
88–95 mol% CO_2	Variable mol ratio (constant volume ratio: 85 + 15)	+5.5
53–74 mol% CO_2	Variable mol ratio (constant volume ratio: 45 + 55)	−3.7

CO_2 phase) and condensing (i.e., dissolution of CO_2 in decane resulting in a sharp decrease in the density of the decane-rich phase compared to the density of a pure decane phase) drive mechanisms, when experimental pressure approaches close to the MMP (i.e., close to the critical pressure) of the system.

Hence, it can be concluded that the experimental data collected in the VIT method can also provide direct and quantitative evidence of prevailing drive mechanism(s) that might be responsible for the development of miscibility in the CO_2-reservoir oil system under investigation. This is an inherent advantage of the VIT method over other experimental methods. Also, the use of experimental input data may avoid the need for the use of any other criteria (e.g., using an arbitrary IFT cutoff [e.g., 0.1–0.3 mN/m] as suggested by Al Riyami and Rao 2015), if an IFT-based calculation procedure is used for estimating an apparent MMP (i.e., condition of near miscibility signified the achievement of low IFT at a pressure close to the MMP).

References

Al Riyami M, Rao DN (2015) Estimation of near-miscibility conditions based on gas-oil interfacial tension calculations. Soc Pet Eng. https://doi.org/10.2118/174646-MS

Adekunle O, Hoffman BT (2016) Experimental and analytical methods to determine minimum miscibility pressure (MMP) for Bakken formation crude oil. J Pet Sci Eng 146:170–182

Ahmad W, Vakili-Nezhaad G, Al-Bemani AS et al (2016) Uniqueness, repeatability analysis and comparative evaluation of experimentally determined MMPs. J Petro Sci Eng 147:218–227

Ayirala SC (2005) Measurement and modeling of fluid-fluid miscibility in multicomponent hydrocarbon systems. Thesis, Louisiana State University

Ayirala SC, Rao DN (2006) Comparative evaluation of a new MMP determination technique. Soc Pet Eng. https://doi.org/10.2118/99606-MS

Ayirala SC, Rao DN (2011) Comparative evaluation of a new gas/oil miscibility-determination technique. Soc Pet Eng. https://doi.org/10.2118/99606-PA

Awari-Yusuf IO (2013) Measurement of crude oil interfacial tension to determine minimum miscibility in carbon dioxide and nitrogen. Thesis, Dalhousie University
Bon J, Sarma HK, Theophilos AM (2005) An investigation of minimum miscibility pressure for
   CO_2-rich injection gases with pentanes-plus fraction. Soc Pet Eng. https://doi.org/10.2118/9753
   6-ms
Chen H, Tang H, Zhang X et al (2017) Decreasing in pressure interval of near-miscible flooding by
   adding intermediate hydrocarbon components. Geo Eng, 21(3):151-157
Clark NJ, Shearin HM, Schultz WP et al. (1958) Miscible drive—its theory and application. Soc
   Pet Eng. https://doi.org/10.2118/1036-g
Christiansen RL, Kim H (1986) Apparatus and method for determining the minimum miscibility
   pressure of a gas in a liquid. US Patent 4,627,273A, 9 Dec 1986
Christiansen RL, Haines HK (1987) Rapid measurement of minimum miscibility pressure with the
   rising-bubble apparatus. Soc Pet Eng. https://doi.org/10.2118/13114-pa
Dong M, Huang S, Srivastava R (2000) Effect of solution gas in oil on CO_2 minimum miscibility
   pressure. Pet Soc Can. https://doi.org/10.2118/00-11-05
Ekundayo JM, Ghedan SG (2013) Minimum miscibility pressure measurement with slim tube
   apparatus—how unique is the value? Soc Pet Eng. https://doi.org/10.2118/165966-ms
Elsharkawy AM, Poettmann FH, Christiansen RL (1996) Measuring CO_2 minimum miscibility
   pressures: slim-tube or rising-bubble method? Energy Fuels 10(2):443–449
Hsu JJC, Nagarajan N, Robinson RL (1985) Equilibrium phase compositions, phase densities,
   and interfacial tension for CO_2 +hydrocarbon systems. 1. CO_2 +n-butane. J Chem Eng Data
   30:485–491
Jessen K, Orr FM (2008) On interfacial-tension measurements to estimate minimum miscibility
   pressures. Soc Pet Eng. https://doi.org/10.2118/110725-pa
Johns RT, Sah P, Solano R (2002) Effect of dispersion on local displacement efficiency for multicomponent enriched-gas floods above the minimum miscibility enrichment. Soc Pet Eng. https://
   doi.org/10.2118/75806-PA
Li S, Luo P (2017) Experimental and simulation determination of minimum miscibility pressure
   for a Bakken tight oil and different injection gases. Petroleum 3(1):79–86
Lim FH, Munoz E, Joshi NB (2008) Design and initial results of EOR and flow assurance lab tests
   for K2 Field development in the deepwater Gulf of Mexico. In: Offshore Technology Conference.
   https://doi.org/10.4043/19624-ms
Mogensen K, Hood P, Lindeloff N et al (2009) Minimum miscibility pressure investigations for a
   gas-injection EOR project in Al Shaheen Field, Offshore Qatar. Soc Pet Eng. https://doi.org/10.
   2118/124109-ms
Nagarajan N, Gasem KAM, Robinson RL (1990) Equilibrium phase compositions, phase densities,
   and interfacial tensions for carbon dioxide +hydrocarbon systems. 6. Carbon dioxide + n-butane +
   n-decane. J Chem Eng Data 35:228–231
Nagarajan N, Robinson RL (1986) Equilibrium phase compositions, phase densities, and interfacial
   tensions for CO_2 +hydrocarbon systems. 2. CO_2 +n-decane. J Chem Eng Data 31:168–171
Novosad Z, Sibbald LR, Costain TG (1990) Design of miscible solvents for a rich gas drive-
   comparison of slim tube and rising bubble tests. Pet Soc Canada. https://doi.org/10.2118/90-01-
   03
Nobakht M, Moghadam S, Gu Y (2008) Determination of CO_2 minimum miscibility pressure from
   measured and predicted equilibrium interfacial tensions. Ind Eng Chem Res 47:8918–8925
Orr FM, Jessen K (2007) An analysis of the vanishing interfacial tension technique for determination
   of minimum miscibility pressure. Fluid Phase Equilib 255:99–109
Orr FM, Silva MK, Lien CL et al (1982) Laboratory experiments to evaluate field prospects for
   CO_2 flooding. Soc Pet Eng. https://doi.org/10.2118/9534-pa
Patil S, Dandekar A, Khataniar S (2008) Phase behavior, solid organic precipitation, and mobility
   characterization studies in support of enhanced heavy oil recovery on the Alaska North Slope.
   Final Report DOE Award no. DE-FC26-01NT41248. United States Department of Energy

Rao DN (1997) A new technique of vanishing interfacial tension for miscibility determination. Fluid Phase Equilib 139(1–2):311–324

Rathmell JJ, Stalkup FI, Hassinger RC (1971) A laboratory investigation of miscible displacement by carbon dioxide. Soc Pet Eng. https://doi.org/10.2118/3483-ms

Saini D (2016a) An investigation of the robustness of physical and numerical vanishing interfacial tension experimentation in determining CO_2 + crude oil minimum miscibility pressure. J Petro Eng 2016:13. https://doi.org/10.1155/2016/8150752

Saini D (2016b) Modeling of pressure dependence of interfacial tension behaviors of a complex supercritical CO2 + live crude oil system using a basic Parachor expression. J Pet Environ Biotechnol 7:277

Saini D, Rao DN (2010) Experimental determination of minimum miscibility pressure (MMP) by gas/oil IFT measurements for a gas injection EOR project. Soc Pet Eng. https://doi.org/10.2118/132389-ms

Sayegh S, Huang S, Zhang YP et al (2007) Effect of H2S and pressure depletion on the CO_2 MMP of Zama oils. Pet Soc Can. https://doi.org/10.2118/07-08-03

Schechter DS, Guo B (1998) Parachors based on modern physics and their uses in IFT prediction of reservoir fluids. Soc Pet Eng. https://doi.org/10.2118/30785-pa

Sequeira DS (2006) Compositional effects on gas-oil interfacial tension and miscibility at reservoir conditions. Thesis, Louisiana State University

Sequeira DS, Ayirala SC, Rao DN (2008) Reservoir condition measurements of compositional effects on gas-oil interfacial tension and miscibility. Soc Pet Eng. https://doi.org/10.2118/113333-ms

Srivastava RK, Huang SS, Dong M (2000) Laboratory investigation of Weyburn CO_2 miscible flooding. Pet Soc Can. https://doi.org/10.2118/00-02-04

Thomas FB, Zhou XL, Bennion DB et al (1994b) A comparative study of RBA, P-x, Multicontact and Slim Tube results. Pet Soc Can. https://doi.org/10.2118/94-02-02

Torabi F, Asghari K (2010) Effect of connate water saturation, oil viscosity and matrix permeability on rate of gravity drainage during immiscible and miscible displacement tests in matrix-fracture experimental model. Soc Pet Eng. https://doi.org/10.2118/141295-pa

Voon C, Awang M (2015) Comparison of MMP between slim tube test and vanishing interfacial tension test. In: Awang M, Negash B, Md Akhir N et al (eds) ICIPEG 2014. Springer, Singapore

Vulin D, Gaćina M, Biličić V (2018). Slim-tube simulation model for CO_2 injection EOR. Rudarsko-geološko-naftni zbornik (Mining-Geological-Petroleum Engineering Bulletin) 33(2). https://doi.org/10.17794/rgn.2018.2.4

Wang Y, Orr FM (1997) Analytical calculation of minimum miscibility pressure. Fluid Phase Equilib 139:101–124

Whorton LP, Brownscombe ER, Dyes AB (1952) Method for producing oil by means of carbon dioxide. US Patent 623,596, 30 Dec 1952

Yarborough L, Smith LR (1970) Solvent and driving gas compositions for miscible slug displacement. Soc Pet Eng. https://doi.org/10.2118/2543-pa

Yellig WF, Metcalfe RS (1980) Determination and prediction of CO_2 minimum miscibility pressures (includes associated paper 8876). Soc Pet Eng. https://doi.org/10.2118/7477-pa

Yuan H, Chopra AK, Marwah V (2008) Fluid characterization for miscible gas floods. Soc Pet Eng. https://doi.org/10.2118/114913-ms

Zhang K (2016) Qualitative and quantitative technical criteria for determining the minimum miscibility pressures from four experimental methods. Thesis, University of Regina

Zhou D, Orr FM (1998) An analysis of rising bubble experiments to determine minimum miscibility pressures. Soc Pet Eng. https://doi.org/10.2118/30786-pa

Chapter 4
Non-experimental Approaches

Abstract Over time, various non-experimental approaches have been developed for predicting the MMP in a fast, reliable, easy-to-use, and inexpensive manner. Some recent approaches are presented below, wherein their capabilities and limitations for accurately determining the MMP for a variety of CO_2-reservoir oil systems are also presented and discussed.

4.1 Introduction

Compared to experimental approaches, which are costly and time consuming and involve tedious experimental procedures, non-experimental approaches provide a relatively simple, faster, and inexpensive way for determining the MMP over a wide range of conditions. These approaches include empirical correlations, analytical methods, and computational or numerical methods.

Apart from experimental characterization and the determination of CO_2-reservoir oil miscibility in terms of the MMP, researchers have also attempted to characterize the phenomenon of CO_2-reservoir oil miscibility by fitting experimentally derived MMP data using a variety of empirical correlations. In contrast, both analytical and numerical methods use well-characterized EOR fluid phase behavior data (i.e., well-tuned EOS models) for calculating the MMP.

The analytical gas-flooding theory, which was developed to mathematically describe the processes of multicontact miscible (MCM) displacement of reservoir oils by injected CO_2, form the basis of deriving analytical solutions for calculating the MMP. One analytical solution that is commonly used in determining the MMP analytically is the method of characteristics (MOC)-based key-tie-line approach. However, the computational or numerical methods either rely on compositional simulation approach for performing one-dimensional (1D) slim-tube simulations or use a single- and multiple-mixing-cell methodology for determining the MMP numerically.

The empirical methods/correlations are often used for estimating the MMP when detailed information is not readily available on an actual CO_2-reservoir oil system.

D. Saini, *CO2-Reservoir Oil Miscibility*, SpringerBriefs in Petroleum Geoscience & Engineering, https://doi.org/10.1007/978-3-319-95546-9_4

They are routinely used for assessing the viability of pure CO_2-EOR or simultaneous CO_2-EOR and storage projects during preliminary feasibility studies. For example, the Advanced Resources International (ARI) reported the use of Cronquist correlation (Mungan 1981) for calculating the MMP while assessing the future CO_2-EOR potential in the large oil fields of the Permian Basin, located in the west part of Texas and New Mexico, USA (ARI 2006). Similarly, Kuuskraa et al. (2011), while building a national CO_2-EOR resource assessment for enhanced energy security via CO_2-EOR and lower CO_2 emissions via CO_2-storage offered by CO_2-EOR operations, used empirically derived MMP values for screening a large number of candidate reservoirs.

The analytical gas-flooding theory allow us to theoretically treat both the phase behavior (compositional path) and flow (high local displacement efficiency [typically >90%]) aspects of MCM displacement of reservoir oils by injected CO_2. In turn, the theory also results in rigorous calculations of the MMP while characterizing the physical mechanism(s) responsible for miscible displacement. Because the MMP calculated from the key-tie-line approach is dispersion free, it is commonly referred to as dispersion-free thermodynamic MMP. The theory of gas injection processes has been discussed in detail by Orr (2007).

The analytical solutions for the flow problem (i.e., physical dispersion-free [piston-like]) displacement of reservoir oil by injected CO_2 also often serve as a theoretical basis to evaluate the robustness of any new MMP determination method or to explain and better understand the underlying physical mechanism(s) responsible for the development of CO_2-reservoir oil miscibility. For example, Orr et al. (1993) used the analytical theory to demonstrate that the assumption of neglecting the effect of the methane content of the oil while developing an empirical correlation is indeed a theoretically sound practice. Jessen and Orr (2008) used the key-tie-line approach (Jessen and Stenby 2007) for predicting the MMP values to demonstrate the accuracy of their EOS model before using it to test the robustness of the VIT method.

Another category of the analytical MMP methods includes the use of Parachor-based basic expressions and the mechanistic Parachor models to calculate the pressure dependence of the IFT behaviors of CO_2-reservoir oil systems. Similar to the VIT method, the calculated IFT versus pressure curve is then extrapolated to the pressure condition of zero IFT (i.e., MMP for a given CO_2-reservoir oil system). Because such analytical methods resemble the VIT method, they are collectively referred to here as simulated VIT methods. In the published literature, the use of both types (i.e., the Parachor-based basic expression and mechanistic Parachor model-based) of the simulated VIT methods for determining the MMP for variety of CO_2-reservoir oil systems have been reported. For example, Nobakht et al. (2008) used the simulated VIT method, which was based on the use of a Parachor-based basic expression, for determining the MMP for a CO_2-Weyburn reservoir oil system. Ashrafizadeh and Ghasrodashti (2011) used the mechanistic Parachor model-based simulated VIT method proposed by Ayirala and Rao (2006) for calculating the MMP for different injection gas-reservoir oil mixtures.

Numerical methods, such as the multiple-mixing-cells method (i.e., cell-to-cell simulation) presented by Ahmadi and Johns (2011), are simple, robust, and more reliable than the MOC-based key-tie-line approach. In the multiple-mixing-cell method, MMP calculations are not affected by the number of components in the gas or oil. The multiple mixing-cell method presented by Jaubert et al. (1998) allows the calculation of the MMP for any type (vaporizing, condensing, mixed) of displacement mechanism. Similarly, the mixing-cell method of Ahmadi and Johns (2011) is accurate for all displacement types, including mixed (vaporizing/condensing) drives, and the MMP can be calculated in minutes using an Excel spreadsheet. In the mixing-cell approach, fraction flow calculations are not performed; instead, only the key tie-lines of the displacement are found.

Another numerical approach—namely 1D slim-tube simulation method—uses 1D grid blocks to numerically mimic the displacement process that is expected to occur in a physical slim-tube displacement test. In this method, the MMP is determined similar to the physical experiment; however, it depends on the number of grid blocks used in the simulation (Jaubert et al. 1998; Dzulkarnain et al. 2011). According to Dzulkarnain et al. (2011), the use of coarse grids will lead to numerical dispersion, caused by truncation error resulting from the approximations of the partial differential equations used for representing the displacement process. Hence, similar to the slim-tube method, the MMP predicted by the 1D slim-tube simulation method is an apparent MMP. On the other hand, if a large number (e.g., 5000) of grids are used, excellent accuracy on the calculated values can be obtained (Jaubert et al. 1998). In this situation, the predicted apparent MMP will be close to thermodynamic MMP.

According to Wang and Peck (2000), 1D slim-tube simulation can predict MMPs that are consistent with the MMPs values obtained from the slim-tube method. However, as shown by Yuan et al. (2008), the apparent MMP determined for a given CO_2-reservoir oil system by 1D slim-tube simulation with a fixed number of grid blocks may be greatly affected by different EOS characterization procedures.

On the other hand, non-experimental approaches have their own inherent limitations. For example, the development of any empirical method itself relies on already available experimental data; hence, the prediction capability and accuracy of an empirical correlation are often limited by the values and ranges of those input parameters observed in the experimental dataset(s) used for building the empirical method. As noted by Yuan et al. (2005), one significant disadvantage of empirical correlations is that they often rely on slim tube data, which are themselves are uncertain. Similarly, the reliability of results obtained from an analytical or numerical method depends on the accuracy of the EOS model used for characterizing the phase behavior of a given reservoir oil and injected CO_2 system. For example, Dzulkarnain et al. (2011) have reported the uncertainty in the MMP estimates which result from the uncertainty in the values of certain EOS model parameters, including relative volume, saturation pressure, and density data.

Reported studies like Mogensen et al. (2009) highlight the role of different non-experimental approaches, which are often verified against experimentally measured MMP data in developing a robust EOS model capable of describing the miscibility and swelling behavior accurately in compositional reservoir simulation studies designed to predict the performance of planned CO_2-EOR projects. Wang and Orr (1997) pointed out that the use of finite grids or numbers of mixing cells in both the numerical and analytical approaches (i.e., 1D slim tube simulations and multiple-mixing-cells methodology, respectively) can result in erroneous MMP values. Similar to a physical slim-tube method, 1D slim-tube simulations require the simulation of several slim-tube type displacements at several pressure steps, which is often cumbersome and time consuming. The MOC-based key-tie-line approach can also lead to incorrect MMP predictions when the tie-line lengths are not monotonic between each successive key tie line (Ahmadi et al. 2011).

According to Nobakht et al. (2008), the simulated VIT method, which uses a Parachor-based basic expression, is generally unsuitable for predicting the equilibrium IFT for CO_2-reservoir oil systems. It is worth mentioning here that any simulated VIT method requires the compositions and densities of the equilibrium liquid (i.e., oil-rich) and vapor (i.e., CO_2-rich) phases for performing IFT calculations. It also requires a prior knowledge of the Parachor, which is believed to be a measure of the molecular volume and chemical composition. In the case of pure substances, it is generally determined from the measured surface tension data. However, for oil-cut (i.e., C_{7+} fraction), Parachor can be calculated using a Parachor correlation. Schechter and Guo (1998) have provided three Parachor correlations. In addition to the above-mentioned input data, the mechanistic Parachor model-based simulated VIT method also requires the ratio of diffusivities between the equilibrium fluid phases. Most of the time, the above-mentioned input data are deduced from "well-tuned" EOS models, which may not converge or cannot perform the flash calculations that are necessary for calculating the desired input data.

Irrespective of certain limitations, non-experimental approaches have greatly enhanced our capabilities to characterize and determine CO_2-reservoir oil miscibility in terms of the MMP. They have also helped us in better understanding the role of both the phase behavior and flow aspects of miscible displacement processes. For example, based on the analytical gas-flooding theory, Orr and Jessen (2007) have emphasized the fact that the combination of flow and phase equilibrium creates the compositions which lead to high-displacement efficiency in a miscible CO_2 flood. On the other hand, based on the mixing-cell approach, Ahmadi and Johns (2011) have supported the notion that the MMP is independent of fractional flow.

Obviously, the non-experimental approaches have evolved over time and are continuing to evolve. For example, via an approach taken by Yuan et al. (2005) (i.e., use of the key-tie-line approach), an empirical correlation to predict the MMP for a wide range of conditions (i.e., reservoir temperature, reservoir oil composition, and injected gas composition) can be calculated quickly and accurately. In some cases, empirical correlations may be less accurate; however, they are quick and easy to use as they rely on only few input parameters that are generally available.

The use of efficient algorithms or statistical modeling techniques have not only resulted in significant reduction in computation time, but it has also improved the accuracy of prediction results obtained using non-experimental approaches. For example, Ahmadi and Johns (2011) used a novel extrapolation technique that resulted in a substantial increase in the speed of the mixing-cell method. Ahmadi et al. (2011) have developed a simple procedure to improve the MOC-based key-tie-line approach.

Some of the latest non-experimental approaches (i.e., empirical correlations and analytical and numerical methods) are discussed next.

4.2 Empirical Approach

There are variety of empirical methods/correlations for determining the MMP for both pure CO_2-reservoir oil and impure CO_2-reservoir oil systems. They are routinely presented and discussed in the published literature (e.g., Emera and Sarma 2005; Rezaei et al. 2013; Zhang et al. 2015; Alomair and Garrouch 2016; Valluri et al. 2017; Mansour et al. 2017). Even though the development of empirical correlations may require experimentally measured MMP data, in the absence of any site-specific experimentally measured MMP, they still offer an effective and convenient means to make initial estimates of the MMP for a given CO_2-reservoir oil system.

As mentioned earlier, majority of empirical correlations use a combination of certain independent parameters, which may include reservoir temperature, molecular weight of the C_{5+} or C_{7+} fractions present in the reservoir oil, fraction of volatile (C_1 and N_2) and intermediate (C_2 through C_6, H_2S, and CO_2) present in reservoir oil, for predicting the MMP for pure CO_2 and reservoir oil system. They can also be extended for impure CO_2 streams, if an injection gas critical property function is included in a given correlation (Alomair and Garrouch 2016).

The empirical correlations employ different statistical technique(s) or mathematical method(s)/scheme(s) for deriving the functional relationship between the MMP and the independent variables. For example, early empirical correlations (e.g., Holm and Josendal 1974) used simple techniques (e.g., graphical solution) for deriving a function relationship between the MMP and the variables affecting the MMP. In contrast, the latest empirical correlations (e.g. Karkevandi-Talkhooncheh et al. 2018) employ more sophisticated and complex mathematical models (e.g., radial basis function [RBF] neural network model further assisted with optimization methods such as imperialistic competitive algorithm [ICA]). Still, the accuracy and trustworthiness of any empirical correlation depends on the comprehensiveness of the used experimental dataset and the range of various independent variables.

Some of the recently developed and simple-to-use empirical correlations for both pure and impure CO_2-reservoir oil systems are briefly discussed here.

4.2.1 Emera-Sarma Correlation

Emera and Sarma (2005) used the concept of the genetic algorithm (GA) for developing a GA-based correlation for predicting the MMP. The genetic algorithm, which is a computer-based search procedure, involves a random generation of potential design solutions accompanied by an evaluation and refinement of the solution until stopping criteria are met (Goldberg 1989; Emera and Sarma 2005). According to Emera and Sarma (2005), GA is an artificial intelligence technique that can be used for both the selection of parameters to optimize the performance of a system and the testing and fitting of quantitative models.

Using the GA approach, Emera and Sarma developed the following power-law type correlation model, which is based on reservoir temperature (T_R, °C), molecular weight of C_{5+} fraction (MW_{C5+}), and the ratio between volatile components (C_1 and N_2) and intermediate components (C_2–C_4, H_2S, and CO_2).

$$MMP = 5.0093 \times 10^{-5} \times (1.8T_R + 32)^{1.164} \times (MW_{C5+})^{1.2785}$$
$$\times \left(\frac{Volatiles}{Interm.}\right)^{0.1073} \tag{4.1}$$

Where MMP is the minimum miscibility pressure in MPa and $\frac{Volatiles}{Interm.}$ is the ratio between mole fractions of volatile components (C_1 and N_2) and intermediate components (C_2–C_4, H_2S, and CO_2) present in the reservoir oil. The above-mentioned correlation also includes two modifications for bubble point pressure. They are:

(a) If the bubble point pressure of the reservoir oil in question is less than 0.345 MPa, the following modified correlation is used.

$$MMP = 5.0093 \times 10^{-5} \times (1.8T_R + 32)^{1.164} \times (MW_{C5+})^{1.2785} \tag{4.2}$$

As evident from Eq. 4.2, the ratio between mole fractions of volatile components (C_1 and N_2) and intermediate components (C_2–C_4, H_2S, and CO_2) is not part of the correlation. Obviously, reservoir oils exhibiting low bubble point pressures behave like stock tank oils which contain no dissolved gas.

(b) If the predicted MMP (Eq. 4.1) is less than bubble point pressure of the reservoir oil in question, the bubble point pressure itself will be selected as MMP. In such scenarios, injected CO_2 will be mixed with the free gas that exists below bubble point pressure, which will adversely affect the miscibility process and MMP value.

According to Emera and Sarma (2005), their GA-based correlation gives a more accurate prediction among all other tested correlations when predicting the MMP for both pure CO_2-reservoir oil and impure CO_2-reservoir oil systems.

4.2.2 Li et al. Correlation

Li et al. (2012) presented an improved MMP correlation for both live and dead reservoir oils and pure CO_2 stream. Here, the terms "live reservoir oil" and "dead reservoir oil" are often used to differentiate between the actual reservoir oil, which may have an appreciable amount of dissolved gas, and the stock tank oil, which has lost any of its dissolved gas. Their correlation is expressed as a function of reservoir temperature (T_R, °C), molecular weight of C_{7+} fraction (MW_{C7+}, g/mol), and mole fraction ratio of volatile components (C_1 and N_2) to intermediate components (C_2–C_6, CO_2, and H_2S). The functional form of the correlation reported by Li et al. (2012) is given by:

$$MMP = 7.30991 \times 10^{-5} \times [\ln(1.8T_R + 32)]^{5.33647} \times [\ln(MW_{C7+})]^{2.08836}$$

$$\times \left(1 + \frac{x_{VOL}}{x_{INT}}\right)^{0.201658} \tag{4.3}$$

Where MMP is the minimum miscibility pressure in MPa, x_{VOL} is the fraction of volatiles in the reservoir oil including C_1 and N_2, mol%, and x_{INT} is the fraction of intermediates in the reservoir oil including C_2–C_6, H_2S, and CO_2, mol%.

According to Li et al. (2012), their correlation, in comparison to other commonly used correlations in the published literature such as Lee (1979), Yellig and Metcalfe (1980), Orr and Jensen (1984), Glaso (1985), Alston et al. (1985), Yuan et al. (2005), Emera and Sarma (2005), Shokir (2007), is most suitable correlation for estimating the MMP for pure CO_2 and various live and dead reservoir oils. It also reveals the modest impact of molecular weight of C_{7+} fraction on the MMP, whereas most of the other correlations that take the molecular weight of C_{5+} or C_{7+} fraction into account overemphasize the contribution of the molecular weight of C_{5+} and C_{7+} fraction on the MMP.

The Li et al. correlation also corrects an inconsistency which can be observed in a correlation such as the Emera-Sarma correlation, if used for determining the MMP for a CO_2-dead reservoir oil system. The dead reservoir oils or in other words the stock tank oils contain no dissolved gas. It means the ratio of volatile to intermediate components should have a zero value. If that is the case, then the original Emera-Sarma correlation (Eq. 4.1) would result in a zero MMP for a CO_2-dead reservoir oil system, which is physically incorrect from an experimental point of view (Li et al. 2012). As demonstrated by Li et al. (2012), the MMP for a CO_2-dead reservoir oil system can be physically measured using an experimental approach such as the slim-tube method.

4.2.3 Zhang et al. Correlation

Taking advantage of the 210 groups of CO_2-reservoir oil MMP experimental data tested by the slim-tube method in the published literature, Zhang et al. (2015) built a uniform empirical correlation to calculate the MMP with four parameters by using the modified conjugate gradient and global optimization algorithm regression theory. The four parameters include reservoir temperature (T_R, °C), relative molecular weight of C_{7+} fraction (MW_{C7+}, g/mol), mole fractions of volatile components (C_1 and N_2)

and mole fractions of intermediate components (C_2–C_6, CO_2, and H_2S) present in the reservoir oil. The correlation can be written as:

$$MMP_{pure} = 8.3397 \times 10^{-5} \times [\ln(1.8T_R + 32)]^{3.9774} \times [\ln(MW_{C7+})]^{3.3179}$$

$$\times \left(1 + \frac{x_{VOL}}{x'_{MED}}\right)^{0.17461} \tag{4.4}$$

Where MMP_{pure} is the minimum miscibility pressure for a given pure CO_2-reservoir oil system in MPa, x_{VOL} is the fraction of volatile components present in the reservoir oil (C_1 and N_2), mol%, and x'_{MED} is the fraction of intermediate components present in the reservoir oil (C_2–C_6, H_2S, and CO_2) mol%.

In comparison to other 11 correlations reported in the published literature—namely Cronquist (1977), Lee (1979), Yellig and Metcalfe (1980), Orr and Jensen (1984), Glaso (1985), Alston et al. (1985), Yuan et al. (2005), Emera and Sarma (2005), Shokir (2007), Ju et al. (2012), and Chen et al. (2013)—the Zhang et al. correlation can be used for a broader range of reservoir temperatures (21.67–191.97 °C), relative molecular weight of C_{7+} fraction (130–402.7 g/mol), and MMP (0–70 MPa).

According to Zhang et al. (2015), the use of MW_{C7+} rather than MW_{C5+} led to a slightly better performance (i.e., high correlation coefficient R^2). They also emphasized the fact that MW_{C7+} is a routine measurement which is performed and reported in a typical compositional analysis study of reservoir oil. They also pointed out the use of logarithm terms for both the reservoir temperature and MW_{C7+} results in reduced influence of these parameters on the MMP, if their values are relatively high. Also, the use of $1 + \frac{x_{VOL}}{x'_{MED}}$ term extends the applicability of the correlation to heavy oils as well because there are fewer volatile components present in heavy oils. It appears that Zhang et al. (2015) used the term "heavy oil" instead of "dead reservoir oil or stock tank oil" because they compared their correlation with the other 11 most popular correlations which were mostly developed for non-heavy (i.e., specific gravity >20°API [American Petroleum Institute]) reservoir oils.

4.2.4 Valluri et al. Correlation

Recently, Valluri et al. (2017) proposed an improved correlation to estimate the MMP for CO_2-reservoir oil systems where simultaneous CO_2-EOR and storage projects for carbon capture, utilization, and storage (CCUS) purposes might be initiated in the future. Their correlation, a power law-based model, is based only on reservoir temperature (T, °F) and molecular weight of the C_{5+} fraction (MW_{C5+}, g/mol). The correlation is:

$$MMP_{pure} = 0.3123 \times T^{0.9851} \times MW_{C5+}^{0.7421} \tag{4.5}$$

Where MMP_{pure} is the minimum miscibility pressure for given pure CO_2-reservoir oil system in MPa. Valluri et al. (2017) used a nonlinear least square solver, which uses the Trust-Region-Reflective optimization approach described in Coleman and Li (1996). According to Valluri et al. (2017), molecular weight of the C_{5+} fraction can be estimated easily from the reservoir oil's specific gravity (i.e., °API) using the following equation:

$$MW_{C5+} = \left(\frac{7864.9}{°API}\right)^{0.9628} \tag{4.6}$$

If information on °API is not easily available, then the molecular weight of C_{7+} fraction (i.e., MW_{C7+}), which is often available, can be used for obtaining MW_{C5+} value using:

$$MW_{C5+} = -0.50602 + 0.9543 MW_{C7+} \tag{4.7}$$

The temperature of the reservoir can be estimated from the regional geothermal gradient, if information on the site-specific reservoir temperature is also not readily available. The use of only two parameters, which may either be already available or can be calculated using other easily available information, for estimating the MMP makes the Valluri et al. correlation a good choice for quick and easy screening of a large number of candidate reservoirs for their suitability for future miscible CO_2 flooding projects.

4.3 Analytical Approach

4.3.1 The MOC-Based Key-Tie-Line Approach

The analytical gas flooding theory (Johns et al. 1993; Johns and Orr 1996) is often used to provide a mathematical description of the processes of MCM displacement of reservoir oils by injected CO_2. According to Wang and Orr (1997), under the assumption of one-dimensional (1D), physical dispersion-free (i.e., piston-like) flow, the analytical solutions of the theory can be obtained by the method of characteristics (MOC) for determining the MMP analytically. The said assumption allows the theoretical treatment of both the phase behavior (compositional path) and flow (high local displacement efficiency [typically >90%]) aspects of MCM displacement of reservoir oils by injected CO_2, which in turn, results in rigorous calculation of the MMP. It is the compositional path (i.e., vaporizing drive, condensing drive, mixed [vaporizing/condensing] drive) that controls the development of multicontact miscibility between the injected CO_2 and reservoir oil, whereas, high local displacement efficiency is largely governed by piston-like flow in 1D mode.

According to Yuan et al. (2005), if an analytical solution of a 1D, dispersion-free flow problem (i.e., displacement of reservoir oil by injected CO_2) is used for determining the MMP, it is essentially an exercise of increasing the pressure and constructing the compositional path in terms of decreasing lengths of the key tie-lines. The MMP is the pressure at which one of the key-tie-line lengths first becomes zero. This approach, which uses the MOC method, is commonly referred to as the key-tie-line approach. The detailed calculation procedure used in the MOC-based key-tine-line approach for determining the dispersion free MMP (i.e., thermodynamic MMP) can be found in the published studies (e.g., Wang and Orr 1997; Jessen et al. 1998; Yuan and Johns 2005). However, salient features of this analytical approach, as summarized in above-mentioned and other published studies (e.g., Monroe et al. 1990; Wang and Peck 2000; Mogensen et al. 2009; Ahmadi and Johns 2011), are given below.

Under the assumption of 1D piston-like displacement, the flow behavior of a general injection gas-reservoir oil system (i.e., a system containing an arbitrary number of components in both the injection gas and reservoir oil) is controlled by a sequence of key tie-lines. In the case of N-component injection gas-reservoir oil system, there exist $N-1$ key tie-lines—namely oil tie-line, gas tie-line, and $N-3$ tie-lines known as crossover tie-lines.

According to Wang and Orr (1997), the oil and gas tie-lines extend through the original oil and injected gas compositions, whereas the remaining n−3 key tie-lines, which are commonly referred as crossover tie-lines, can be identified using the MOC theory as a series of tie-lines whose extensions intersect sequentially in composition space. Essentially, the application of the MOC theory enables the construction of an analytical solution describing the composition path from the initial gas composition, which is the case at the injector well, to the initial oil composition, which is the case at the production well (Mogensen et al. 2009).

A well-tuned EOS model is used for performing the necessary flash calculations needed for determining the length of each key tie-line at a given pressure. The minimum pressure at which the length of any one of these key tie-lines is reduced to zero length (Fig. 4.1) is the MMP of the system. The key tie-line with zero length, which intersects a critical point as a tangent, is denoted as a critical tie-line.

In the MOC-based key-tie-line approach, if the critical tie-line at MMP extends through the initial oil composition, the oil tie-line controls the miscibility and the displacement mechanism is the vaporizing drive. On the other hand, if the critical tie-line at MMP extends through the initial gas composition, the gas tie-line controls the development of miscibility and the displacement is the condensing drive. If any of the crossover tie-lines becomes the critical tie-line, the miscibility is developed through mixed (vaporizing/condensing) drive.

Due to the complex nature of calculations that are performed in the MOC-based key-tie-line approach, researchers have noted the use of software packages such as MI-PVT offered by the Tie-Line Technology, Denmark (Jessen and Orr 2008) and the PVTsim offered by Calsep (Mogensen et al. 2009; Kanatbayev et al. 2015). On the other hand, according to Ahmadi et al. (2011), it is possible that MOC-based key-tie-line approach may result in erroneous MMP predictions for multicomponent

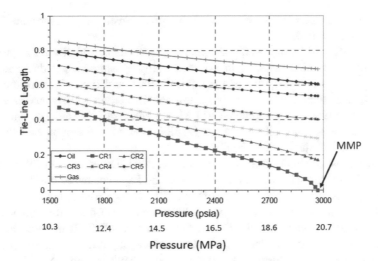

Fig. 4.1 Determination of the MMP from a key-tie-line of zero length, modified figure reproduced by permission: Soc Pet Eng, Fluid characterization for miscible gas floods. Soc Pet Eng. https://doi.org/10.2118/114913-ms, Yuan H, Chopra AK, Marwah V © 2008

gases as the solution may converge to the wrong set of key tie-lines. It is a potential drawback of the approach. A detailed discussion of other limitations of the MOC-based key-tie-line approach and recommendations to further improve its accuracy has also been provided by Ahmadi et al. (2011).

4.3.2 The Simulated VIT Methods

The simulated VIT methods use the calculated pressure dependence of the IFT behaviors of CO_2-reservoir oil systems for obtaining the condition of Zero IFT (i.e., MMP). Nobakht et al. (2008) used a simulated VIT method, which used a Parachor-based basic expression for calculating CO_2-reservoir oil IFT at various pressure steps, for determining the MMP for CO_2 + Weyburn reservoir oil system. They used the EOS-generated vapor-liquid equilibrium (VLE) data for obtaining the IFT versus pressure curve. The curve was then extrapolated to a zero IFT condition for obtaining the MMP. A comparison of the simulated VIT method's results with previously published results revealed that the simulated VIT method always significantly overestimated the MMP for a CO_2 + Weyburn reservoir oil system.

Ashrafizadeh and Ghasrodashti (2011) used the simulated VIT method, which was based on the mechanistic Parachor model developed by Ayirala and Rao (2004), to estimate the MMP for CO_2-Zick's oil 5 system. They used five different EOS models for generating the necessary VLE data needed for calculating the IFT versus pressure curve. The simulated MMPs were found be in good agreement with measured MMP (the slim-tube method).

Recently, Saini (2016a) used a simulated VIT method, which used a Parachor-based basic expression and certain experimentally-measured input data including the compositions and densities of the equilibrium liquid (i.e., oil-rich) and vapor (i.e., CO_2-rich) phases, for determining the MMPs certain CO_2-reservoir oil systems. When compared, the simulated VIT MMPs calculated by Saini (2016a) agreed well with the experimentally-measured MMPs reported in the published literature (Sequeira 2006; Sequeira et al. 2008).

Some published studies (e.g., Orr and Jessen 2007; Jessen and Orr 2008), which used the simulated VIT method that was based on Parachor-based basic expression for modeling the pressure dependence of the IFT behaviors of complex CO_2-reservoir oil systems, have reported large deviations between the VIT method-estimates of the MMP (i.e., simulated VIT method) and the slim-tube MMP. It is noted here that such studies relied on calculated (i.e., EOS-derived) input data (i.e., compositions and densities of the equilibrium CO_2-rich and oil-rich phases and molecular weights of various components present in the system) This observed discrepancy between the results of simulated VIT method and the slim-tube method appears to arise from certain common limitation of cubic EOSs.

As highlighted by Ashrafizadeh and Ghasrodashti (2011), cubic EOSs such as Peng-Robinson EOS (PREOS) are either not able to perform flash calculations for certain CO_2-reservoir oil mixtures or they are not able to predict the equilibrium fluid phase densities in the near-critical region. The failure of PREOS to accurately predict compositions of different phases and yield poor liquid densities even for one-component fluids has already been well documented in the published literature (e.g., Sahimi et al. 1985). Hence, significant error may occur in the predicted MMP values if a simulated VIT method, which employs a cubic EOS-derived input data instead of experimentally-derived input data, models the pressure dependence of the IFT behaviors of complex CO_2-reservoir oil systems.

In their analyses of VIT experiments for ternary and quaternary systems using the EOS-coupled basic Parachor expression, Orr and Jessen (2007) were intrigued by a reported measurement of IFT (Ayirala and Rao 2006) at pressures above the bubble point pressure of a CO_2-hydrocarbon mixture for which simulated VIT method resulted in the termination of IFT versus pressure curve at the predicted bubble point pressure of the feed mixture. Unlike the calculated IFT versus pressure curve that may be terminated at the bubble point/dew point pressures of the feed mixture compositions, experimentally-measured IFT versus pressure curve can be obtained via a pressure which may be well below the bubble point/dew point pressure of a given feed mixture (e.g., Fig. 24 by Ayirala and Rao 2006). As suggested by Ashrafizadeh and Ghasrodashti (2011), the formation of saturated mixtures makes it impossible to perform flash calculations and results in the termination of calculations at saturation pressures.

In contrast to the simulated VIT method, experimental IFT measurements can be performed for any chosen feed mixture composition at any experimental pressure step. Hence, it is not the VIT method that makes erroneous MMP predictions, but the use of an EOS-coupled simulated VIT method leads to a faulty conclusion about the applicability of the VIT method in characterizing and determining the CO_2-reservoir oil miscibility. Also, the analysis of reported experimental and the calculated pressure dependence of the IFT behaviors of complex CO_2-reservoir oil systems, which was presented earlier, clearly demonstrates that it is not necessary to know the so-called "optimal feed mixture composition" beforehand for using the VIT method for determining the MMP.

4.4 Numerical Approach

Numerically, the MMP can be calculated using methods such as the 1D slim-tube simulation and the multiple-mixing-cell method. A detailed discussion of these methods can be found elsewhere (Jaubert et al. 1998, Yuan et al. 2008, Ahmadi and Johns 2011). Both methods are briefly discussed below.

4.4.1 1D Slim-Tube Simulation

The 1D slim-tube simulation method essentially performs numerical slim-tube experiments using a one-dimensional horizontal EOS compositional simulation model via acceptable numbers of grid blocks between the injection and production ends. Similar to a physical slim-tube experiment, in the 1D slim-tube simulation approach, numerically obtained cumulative oil recovery factors at 1.2 cumulative pore volumes injected are plotted against the injection pressure for obtaining the MMP from the intersection point of two trend lines (high slope and low slope) on the cumulative oil recovery versus injection pressure plot (Fig. 4.2, after Vulin et al. 2018).

The main purpose of 1D slim-tube simulation is to validate the ability of a "well-tuned" EOS model to predict the experimentally-measured MMP using an experimental approach before the developed EOS model can be used in other comprehensive compositional simulation modeling studies. Because the 1D slim-tube simulation approach involves the simulation of a dynamic process (i.e., the interplay between the phase behavior and the local displacement efficiency), it is important that the used EOS model is able to appropriately capture both aspects. For this, as suggested by Khabibullin et al. (2017), there needs to be a check on the so-called high vaporization power of the used EOS (i.e., achievement of very high recovery factors at low pressure that may not be confirmed by physical slim-tube tests) in dynamic simulations after necessary tuning of the EOS model using the available physical PVT data.

If the MMP estimated from the 1D slim-tube simulation is still higher than the experimentally-measured MMP, it is necessary to develop a new initial EOS with a different characterization procedure and use the same for achieving a satisfactory match between the simulated MMP and the experimentally-measured MMP and to

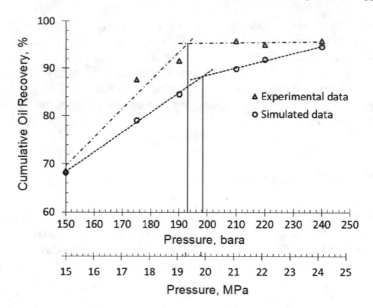

Fig. 4.2 Comparison of experimentally-derived slim-tube MMP with 1D slim-tube simulation based MMP, modified figure reproduced by permission within the scope of Creative Commons 4.0 license: Faculty of Mining, Geology and Petroleum Engineering University of Zagreb, Slim-tube simulation model for CO_2 injection EOR, Rudarsko-geološko-naftni zborni (Mining-Geological-Petroleum Engineering Bulletin). https://doi.org/10.17794/rgn.2018.2.4, Vulin D, Gaćina M, Bilićić V © 2018

confirm that oil recovery is not affected by numerical dispersion (Yuan et al. 2008). Also, as used by Yuan et al. (2008), in the absence of system-specific gas-oil relative permeability curves, published data, which may be available for a given physical slim-tube configuration, can be used.

As emphasized in different published studies (e.g., Yuan et al. 2008; Khabibullin et al. 2017; Wang et al. 2018), to maintain the influence of numerical dispersion on model results within acceptable limits, the number of grid blocks included in the model may vary somewhere between 120 and 2000 for representing a 12–15 m long slim-tube model. As mentioned earlier, numerical dispersion is caused by truncation errors resulting from the approximations of the partial differential equations used for representing the displacement process. Hence, it is necessary that its influence should be kept similar to the magnitude of physical dispersion that may be expected in a physical slim-tube experiment. Another way to compensate for the effect of numerical dispersion on simulation result is to adjust some simulation parameters, such as gas-oil relative permeabilities curves which are needed to perform the simulation. The general shape of gas-oil relative permeability curves at different conditions (i.e., immiscible, near miscible, and miscible) is shown in Fig. 4.3, after Vulin et al. 2018.

The 1D slim-tube simulation approach is considered a slow but safest approach for numerically determining the MMP; however, at times, it may be tedious and

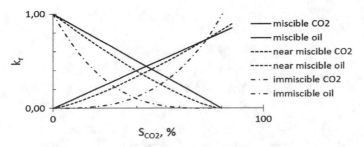

Fig. 4.3 General shape of gas-oil relative permeability curves at miscible, near miscible, and immiscible conditions, figure reproduced by permission within the scope of Creative Commons 4.0 license: Faculty of Mining, Geology and Petroleum Engineering University of Zagreb, Slim-tube simulation model for CO_2 injection EOR, Rudarsko-geološko-naftni zbornik (Mining-Geological-Petroleum Engineering Bulletin). https://doi.org/10.17794/rgn.2018.2.4, Vulin D, Gaćina M, Biličić V © 2018

time consuming. An alternative is another numerical method—namely the multiple-mixing cell method, which only relies on the phase behavior aspect of the MCM process, can be used. The same is briefly discussed next.

4.4.2 The Multiple-Mixing-Cell Method

As both Jaubert et al. (1998), Ahmadi and Johns (2011) have claimed, the multiple-mixing-cell approach, which relies on performing the pressure-temperature (P/T) flash calculations using any well-tuned EOS model, is a simple, practical, and robust numerical method for calculating the MMP. Unlike other two numerical methods—namely the MOC-based-key-tie-line and 1D slim-tube simulation methods, no displacement related calculations are required in the multiple-mixing-cell method. Because only thermodynamics equations are used, the multiple-mixing-cell approach is faster and less cumbersome than the 1D slim-tube simulation method and the accuracy of the results is comparable to both the MOC-based-key-tie-line and 1D slim-tube simulation methods. The multiple-mixing-cell method can predict the MMP for any type of displacement mechanism(s) (i.e., either vaporizing or condensing or mixed) that might be the prevailing physical mechanism in an MCM displacement process.

The method presented by Ahmadi and Johns (2011) uses a variable number of cells for performing the repeated contacts that start with two cells filled with injection gas (G) and reservoir oil (O). As can be seen from Fig. 4.4, after first contact mixing, flashing of the resulting overall composition using the tuned EOS model, results in two new equilibrium compositions—the equilibrium liquid composition (X) and the equilibrium gas composition (Y). Both equilibrium compositions (i.e., X and Y) are mixed again. The mixing is done assuming that the gas moves ahead of the oil phase. Each of the repeated contacts (i.e., 2nd, 3rd, and so on) results in new equilibriums

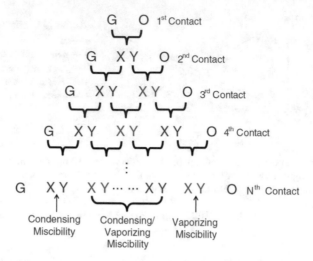

Fig. 4.4 Schematic depiction of repeated contacts in the multiple-mixing-cell method (G: injecting as composition, O: oil composition, Y: equilibrium gas composition, Y: equilibrium liquid composition), reprinted by permission: Soc Pet Eng, Multiple-mixing-cell method for MMP calculations, https://doi.org/10.2118/116823-pa, Ahmadi K, Johns RT © 2011

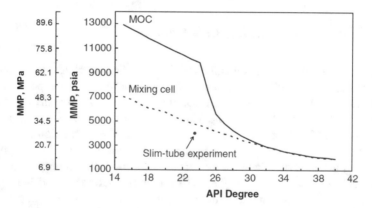

Fig. 4.5 Comparison of predicted MMPs using the MOC and multiple-mixing-cell methods for compositional variations in the Al-Shaheen oil (Mogensen et al. 2009), reprinted by permission: Soc Pet Eng, Limitations of current method-of-characteristics (MOC) methods using shock-jump approximations to predict MMPs for complex gas/oil displacements, https://doi.org/10.2118/1297 09-pa, Ahmadi K, Johns RT, Mogensen K et al. © 2011

compositions which are used in the next set of contacts. The process is continued until all $N_c - 1$ key tie-lines develop and converge to a specific tolerance after N contacts. The MMP is the pressure at which any of the key tie-line first becomes zero length, however in practice, the smallest key tie-line is extrapolated to zero length for estimating the MMP. It is noted that N_c is the number of components present in a given CO_2-reservoir oil system. Also, if the process is repeated for N contacts, a total $2N+2$ cells will be involved.

A detailed procedure for the multiple-mixing-cell method presented by Ahmadi and Johns can be found elsewhere (Ahmadi and Johns 2011). However, as they emphasized, their method always results in finding a single unique set of key-tie-lines for a displacement. On the other hands, in the case of the MOC-based-key-tie-line approach, a check of the tie-line lengths for ensuring the convergence to the correct set of key-tie-lines is often required. Also, their method automatically corrects for numerical dispersion by performing additional contacts, whereas other mixing-cell approaches (e.g., Jaubert et al. 1998), as noted by Ahmadi and Johns (2011), may be susceptible to numerical dispersion due to the use of a fixed number of cells.

It is worth mentioning here that both the analytical (i.e., MOC-based key-tie-line) and the numerical (i.e., multiple-mixing-cell) approaches generally agree well, however, for some CO_2-reservoir oil systems, the predicted MMPs do not match the predictions of Ahmadi and Johns (2011). For example, the difference between the predicted MMP values for the pure CO_2–Al Shaheen reservoir oil system obtained from both methods continued to grow as the API of the oil decreased (Fig. 4.5).

On the other hand, the multiple-mixing-cell approach provide a simple, robust, and faster alternative to the conventional 1D slim-tube simulation approach, when it comes to accurate numerical determination of the MMP for any CO_2-reservoir oil system without worrying about the number of components in the injection gas and reservoir fluid or the displacement mechanism(s).

References

Ahmadi K, Johns RT (2011) Multiple-mixing-cell method for MMP calculations. Soc Pet Eng. https://doi.org/10.2118/116823-PA

Ahmadi K, Johns RT, Mogensen K et al (2011) Limitations of current method-of-characteristics (MOC) methods using shock-jump approximations to predict MMPs for complex gas/oil displacements. Soc Pet Eng. https://doi.org/10.2118/129709-PA

Alomair OA, Garrouch AA (2016) A general regression neural network model offers reliable prediction of CO_2 minimum miscibility pressure. J Petrol Explor Prod Technol 6:351

Alston RB, Kokolis GP, James CF (1985) CO_2 minimum miscibility pressure: a correlation for impure CO_2 streams and live oil systems. SPE J 25(2):268–274

ARI (2006) Basin oriented strategies for CO_2 enhanced oil recovery. Permian Basin, U.S. Department of Energy. http://www.adv-res.com/pdf/Basin%20Oriented%20Strategies%20-%20Permian_Basin.pdf. Accessed 6 May 2018

Ashrafizadeh SN, Ghasrodashti AA (2011) An investigation on the applicability of parachor model for the prediction of MMP using five equations of state. Chem Eng Res Des 89(6):690–696

Ayirala SC, Rao DN (2004) Application of a new mechanistic parachor model to predict dynamic gas-oil miscibility in reservoir crude oil-solvent systems. Soc Pet Eng. https://doi.org/10.2118/91920-MS

Ayirala SC, Rao DN (2006) Comparative evaluation of a new MMP determination technique. Soc Pet Eng. https://doi.org/10.2118/99606-MS

Chen BL, Huang HD, Zhang Y et al (2013) An improved predicting model for minimum miscibility pressure (MMP) of CO_2 and crude oil. J Oil Gas Tech 35(2):126–130

Coleman TF, Li YY (1996) An interior trust-region approach for nonlinear minimization subject to bounds. SIAM J Optim 6:418–445

Dzulkarnain I, Awang MB, Mohamad AM (2011) Uncertainty in MMP prediction from EOS fluid characterization. Soc Pet Eng. https://doi.org/10.2118/144405-MS

Emera MK, Sarma HK (2005) Use of genetic algorithm to estimate CO_2–oil minimum miscibility pressure—a key parameter in design of CO_2 miscible flood. J Pet Sci Eng 46(1–2):37–52

Glaso O (1985) Generalized minimum miscibility pressure correlation (includes associated papers 15845 and 16287). Soc Pet Eng. https://doi.org/10.2118/12893-PA

Goldberg DE (1989) Genetic algorithms in search, optimisation, and machine learning. Addison-Wesley Publishing, USA

Holm LW, Josendal VA (1974) Mechanisms of oil displacement by carbon dioxide. Soc Pet Eng. https://doi.org/10.2118/4736-PA

Jaubert J-N, Wolff L, Neau E et al (1998) A very simple multiple mixing cell calculation to compute the minimum miscibility pressure whatever the displacement mechanism. Ind Eng Chem Res 37(12):4854–4859

Jessen K, Orr FM (2008) On interfacial-tension measurements to estimate minimum miscibility pressures. Soc Pet Eng. https://doi.org/10.2118/110725-PA

Jessen K, Stenby EH (2007) Fluid characterization for miscible EOR projects and CO_2 sequestration. Soc Pet Eng. https://doi.org/10.2118/97192-PA

Jessen K, Michelsen ML, Stenby EH (1998) Global approach for calculation of minimum miscibility pressure. Fluid Phase Equilib 153:251

Johns RT, Orr FM (1996) Miscible gas displacement of multicomponent oils. Soc Pet Eng. https://doi.org/10.2118/30798-PA

Johns RT, Dindoruk B, Orr FM (1993) Analytical theory of combined condensing/vaporizing gas drives. Soc Pet Eng. https://doi.org/10.2118/24112-PA

Ju BS, Qin JS, Li ZP et al (2012) A prediction model for the minimum miscibility pressure of the CO_2-crude oil system. Acta Petrolei Sinica 33(2):274–277

Kanatbayev M, Meisingset K, Uleberg K (2015) Comparison of MMP estimation methods with proposed workflow. Soc Pet Eng. https://doi.org/10.2118/173827-MS

Karkevandi-Talkhooncheh A, Rostami A, Hemmati-Sarapardeh A et al (2018) Modeling minimum miscibility pressure during pure and impure CO_2 flooding using hybrid of radial basis function neural network and evolutionary techniques. Fuel 220:270–282

Khabibullin R, Emadi A, Grin A et al (2017) Investigation of CO_2 application for enhanced oil recovery in a North African field—a new approach to EOS development. EAGE. https://doi.org/10.3997/2214-4609.201700276

Kuuskraa VA, Leeuwen TV, Wallace M (2011) Improving domestic energy security and lowering CO_2 emissions with "Next Generation" CO_2-Enhanced Oil Recovery (CO_2-EOR), Energy Sector Planning and Analysis (ESPA). The US Department of Energy. http://www.midwesterngovernors.org/documents/NETL_DOE_Report.pdf. Accessed 6 May 2018

Lee JI (1979) Effectiveness of carbon dioxide displacement under miscible and immiscible conditions. Report RR-40, Petroleum Recovery Institute, Calgary

Li H, Qin Q, Yang D (2012) An improved CO_2-oil minimum miscibility pressure correlation for live and dead crude oils. Ind Eng Chem Res 51(8):3516–3523

Mansour EM, Al-Sabagh AM, Desouky SM et al (2017) A new estimating method of minimum miscibility pressure as a key parameter in designing CO_2 gas injection process. Egyptian J Pet. https://doi.org/10.1016/j.ejpe.2017.12.002 (in press)

Mogensen K, Hood P, Lindeloff N et al (2009) Minimum miscibility pressure investigations for a gas-injection EOR project in Al Shaheen field, offshore Qatar. Soc Pet Eng. https://doi.org/10.2118/124109-ms

Monroe WW, Silva MK, Larson LL et al (1990) Composition paths in four-component systems: effect of dissolved methane on 1D CO_2 flood performance. Soc Pet Eng. https://doi.org/10.2118/16712-PA

Mungan N (1981) Carbon dioxide flooding-fundamentals. Pet Soc Can. https://doi.org/10.2118/8 1-01-03

Nobakht M, Moghadam S, Gu Y (2008) Determination of CO_2 minimum miscibility pressure from measured and predicted equilibrium interfacial tensions. Ind Eng Chem Res 47:8918–8925

Orr FM (2007) Theory of gas injection processes. Tie-Line Publications, Denmark

Orr FM, Jensen CM (1984) Interpretation of pressure-composition phase diagrams for CO_2/crude-oil systems. Soc Pet Eng. https://doi.org/10.2118/11125-PA

Orr FM, Jessen K (2007) An analysis of the vanishing interfacial tension technique for determination of minimum miscibility pressure. Fluid Phase Equilib 255(2):99–109

Orr FM, Johns RT, Dindoruk B (1993) Development of miscibility in four-component CO_2 floods. Soc Petr Eng. https://doi.org/10.2118/22637-PA

Rezaei M, Eftekhari M, Schaffie M et al (2013) A CO_2-oil minimum miscibility pressure model based on multi-gene genetic programming. Energy Explor Exploit 31(4):607–622

Saini D (2016a) Modeling of pressure dependence of interfacial tension behaviors of a complex supercritical CO_2 + live crude oil system using a basic Parachor expression. J Pet Environ Biotechnol 7:277

Sahimi M, Davis HT, Scriven LE (1985) Thermodynamic modeling of phase and tension behavior of CO_2/hydrocarbon systems. Soc Petr Eng. https://doi.org/10.2118/10268-PA

Schechter DS, Guo B (1998) Parachors based on modern physics and their uses in IFT prediction of reservoir fluids. Soc Pet Eng. https://doi.org/10.2118/30785-PA

Sequeira DS (2006) Compositional effects on gas-oil interfacial tension and miscibility at reservoir conditions. Thesis, Louisiana State University

Sequeira DS, Ayirala SC, Rao DN (2008) Reservoir condition measurements of compositional effects on gas-oil interfacial tension and miscibility. Soc Pet Eng. https://doi.org/10.2118/11333 3-MS

Shokir EM (2007) CO_2-oil minimum miscibility pressure model for impure and pure CO_2 streams. J Pet Sci Eng 58(1–2):173–185

Valluri MK, Mishra S, Schuetter J (2017) An improved correlation to estimate the minimum miscibility pressure of CO_2 in crude oils for carbon capture, utilization, and storage projects. J Pet Sci Eng 158:408–415

Vulin D, Gaćina M, Biličić V (2018) Slim-tube simulation model for CO_2 injection EOR. In: Rudarsko-geološko-naftni zbornik (Mining-Geological-Petroleum Engineering Bulletin), vol 33, no 2. https://doi.org/10.17794/rgn.2018.2.4

Wang Y, Orr FM (1997) Analytical calculation of minimum miscibility pressure. Fluid Phase Equilib 139:101–124

Wang Y, Peck DG (2000) Analytical calculation of minimum miscibility pressure: comprehensive testing and its application in a quantitative analysis of the effect of numerical dispersion for different miscibility development mechanisms. Soc Pet Eng. https://doi.org/10.2118/59378-MS

Wang G, Pickup GE, Sorbie KS et al (2018) The analysis of compositional effects on global flow regimes in CO_2 near-miscible displacements in heterogeneous systems. Soc Pet Eng. https://doi.org/10.2118/190273-MS

Yellig WF, Metcalfe RS (1980) Determination and prediction of CO_2 minimum miscibility pressures (includes associated paper 8876). Soc Pet Eng. https://doi.org/10.2118/7477-PA

Yuan H, Johns RT (2005) Simplified method for calculation of minimum miscibility pressure or enrichment. Soc Pet Eng. https://doi.org/10.2118/77381-PA

Yuan H, Chopra AK, Marwah V (2008) Fluid characterization for miscible gas floods. Soc Pet Eng. https://doi.org/10.2118/114913-MS

Yuan H, Johns RT, Egwuenu AM et al (2005) Improved MMP correlation for CO_2 floods using analytical theory. Soc Pet Eng. https://doi.org/10.2118/89359-PA

Zhang H, Hou D, Li K (2015) An improved CO_2-crude oil minimum miscibility pressure correlation. J Chem Article ID 175940. https://doi.org/10.1155/2015/175940

Chapter 5
Recent Advancements

Abstract Both the experimental and non-experimental approaches for characterizing and determining the CO_2-reservoir oil miscibility in terms of the MMP have continued to evolve. The recent advancements that include improvements in existing experimental approaches and the development of new experimental approaches are presented and discussed. An analysis of the extension of existing and the newly developed experimental and non-experimental methods for characterizing and determining the MMP in unconventional (i.e., tight or shale) reservoirs is also presented.

5.1 Introduction

The growing importance of characterizing and determining CO_2-reservoir oil miscibility in a fast, simple, and robust manner is also evident from ongoing efforts to improve upon the existing approaches and to develop new experimental and non-experimental methods. The future potential of both the CO_2-EOR and simultaneous CO_2-EOR and storage projects in unconventional (e.g., tight oil or shale) reservoirs to augment the overall oil recovery factor while permanently storing large amounts of anthropogenic CO_2 has also prompted the research community to characterize and determine CO_2-reservoir oil miscibility for unconventional reservoirs.

Recent advancements in improving existing experimental approaches for determining and characterizing CO_2-reservoir oil miscibility in both the conventional and unconventional reservoirs include a modification to the standard experimental protocol used in the slim-tube method (Mogensen 2016) and the development of a fast-slim tube method (Adel et al. 2016). The novel protocol proposed by Mogensen (2016) may require fewer slim-tube runs while yielding a more accurate estimation of the MMP. In contrast, the fast-slim-tube method proposed by Adel et al. (2016) advocates the use of a short length (6 m) slim-tube for reducing the time required for running slim-tube experiments without compromising the accuracy of the slim-tube method. A newer version of the VIT method has also been presented recently by Hawthorne et al. (2016) for experimentally determining the MMP for both conventional and unconventional CO_2-reservoir oil systems.

D. Saini, *CO2-Reservoir Oil Miscibility*, SpringerBriefs in Petroleum Geoscience & Engineering, https://doi.org/10.1007/978-3-319-95546-9_5

A new MMP prediction model based on the modified Parachor Model associated with the Perturbed-Chain Statistical Associating Fluid Theory (PC-SAFT) has been developed by Wang et al. (2016) for determining CO_2-MMP both in the bulk phase (i.e., conventional reservoirs) and nanopores (i.e., unconventional reservoirs). Moreover, Saini (2016a) has demonstrated the capability of a basic Parachor model in modeling the experimentally measured VIT MMP. An IFT-based method reported by Al Riyami and Rao (2015), which relies only on using a well-tuned cubic EOS model and the Parachor model, enables rapid calculation of MMP for multicomponent CO_2-reservoir oil systems.

As far as the development of new experimental approaches is concerned, the fast fluorescence-based microfluid MMP method (Nguyen et al. 2015), the sonic response method (SRM), and the rapid pressure increase method (RPI) presented by Czarnota et al. (2017a, b) are among newly developed experimental approaches. These methods are fast, reliable, and more easily conducted than other experimental techniques of MMP measurements.

Keeping the unsuitability of the slim-tube method in mind, as recently suggested by Wang et al. (2017), several recent published studies (e.g., Adekunle and Hoffman 2016; Li and Luo 2017) have attempted to extend the use of one of the existing experimental approaches—the RBA method—for determining the MMP for Bakken formation crude oil. It is worth mentioning that the Bakken formation is one of the major US unconventional tight oil plays.

On the other hand, Zhang et al. (2016) have proposed a revision to existing MMP correlations for their applicability in unconventional tight oil reservoirs. Their integrated MMP correlation, a solubility-based method, appropriately incorporates any of the alternation in MMP value due to nanoscale pore confinement.

Next, some of the above-mentioned recent (i.e., 2015 to present) advancements are briefly described and discussed.

5.2   Improvements in Existing Experimental Approaches

5.2.1   The Novel Slim-Tube Protocol

The modification presented by Mogensen (2016) overcomes the inconsistency in interpretation methodology that is currently used in the slim-tube method. An inconsistency in the interpretation of collected slim-tube displacement experiment (i.e., measured MMP) often arises from the subjective selection of the intersection of the two distinct recovery trend lines or from the selection of a particular cut-off recovery on cumulative oil recovery versus injection pressure plot. Such inconsistency will be more pronounced if no definite intersection point is obtained from two recovery trend lines.

To avoid such inconsistency, Mogensen (2016) proposed tracking the methane-to-propane (C_1/C_3) ratio of the produced gas with respect to the pore volumes injected.

Because of chromatographic separation, a distinct mass transfer signature (i.e., formation of a methane bank ahead of the connected gas front) quantified in terms of C_1/C_3 ratio, can be experimentally observed, especially when flow is immiscible. It is worth mentioning here that the term "chromatographic separation" is often used to characterize relatively fast displacement of the most volatile components such as methane [C_1] by injected CO_2 compared to other hydrocarbon components such as propane [C_3], a scenario which is similar to the separations that may be observed in routine chromatographic analysis of a crude oil sample. The above-mentioned distinct mass transfer signature (i.e., C_1/C_3 ratio) can easily be obtained by performing a compositional analysis of produced gas. Obviously, such a distinct mass transfer signature will either be negligible or completely disappear, if a displacement test is being performed at a pressure that is close to MMP.

According to Mogensen (2016), the proposed protocol would only require four slim-tube displacement experiments and a total of 28 gas sample analyses. Any cost saving, which is likely to be the case due to fewer slim-tube experiments, is expected to cover the additional cost of gas sample analyses. However, it is not clear if such a novel protocol will be equally applicable if no solution gas is present in the reservoir oil sample to begin with. One potential alternative may be the tracking of the ratio of some other less volatile components that may still be extracted by the injected CO_2 and can move ahead of the connected gas front.

5.2.2 The Fast Slim-Tube Method

The main premise of the fast slim-tube method presented by Adel et al. (2016) is that if a short length slime tube, which is sufficient enough to host the mixing zone (as shown in Fig. 1.8) and the velocity of the displacement is slow enough to avoid any viscous fingering is used for running the slim-tube experiment, experimental time can be significantly reduced. In the case of light oil, a slim-tube length of 6 m, which is significantly shorter than typical slim-tube length ranging from 12 m to as high as 24.4 m, is suggested by Adel et al. (2016). They recommend the use of three data points below MMP, and three more above it.

The API gravity of the light oil samples used by Adel et al. (2016) to demonstrate the validity of their method ranged from 34.9° to 42.91°. Hence, they cautioned about the use of this technique in studying CO_2 miscibility in heavier crude oils, as longer slim-tube length will be needed in order to host the mixing zone in such CO_2-reservoir oil systems.

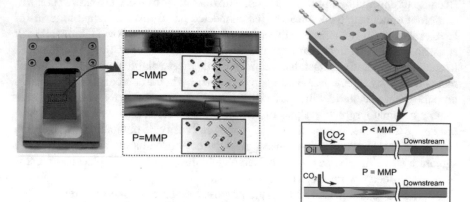

Fig. 5.1 Schematic depiction of the fast fluorescence-based microfluidic experimental setup modified figure reproduced by permission within the scope of Creative Commons 4.0 license: American Chemical Society (ACS), Fast fluorescence-based microfluidic method for measuring minimum miscibility pressure of CO_2 in crude oils, https://pubs.acs.org/doi/10.1021/ac5047856, Nguyen P, Mohaddes D, Riordon J et al. © 2015. Any further permissions related to the material excerpted should be directed to the ACS

5.3   Newly Developed Experimental Approaches

5.3.1   The Fast Fluorescence-Based Microfluidic MMP Method

Recently, Nguyen et al. (2015) reported a rapid, user-independent method—namely the fast fluorescence-based microfluidic MMP method—for measuring CO_2-in-oil MMP at reservoir-relevant temperatures. In this method, inherent fluorescence of reservoir crude oil is used for quantitatively observing the mixing of continuously generated CO_2 bubbles within a reservoir oil sample-filled microfluidic chip embedded in silicon (Fig. 5.1). As can be seen in Fig. 5.2a, experimental pressure below the MMP, CO_2 bubbles retain their shape with sharp blocking interfaces because of high IFT between CO_2 and reservoir oil. As the pressure is increased to MMP, the IFT between CO_2 and reservoir oil decreases which leads to deformation of CO_2 bubbles in response to flow induced stresses (Fig. 5.2b). Above MMP, mixing is rapid, and the two phases are largely indistinguishable at the downstream location shown in Fig. 5.2c.

According to Nguyen et al. (2015), compared to all existing experimental methods, their fluorescence-based microfluidic method directly measures the MMP. For this it measures the fluctuations in fluorescence intensity at experimental pressure over time in an operator-independent manner. An automated fluorescence analysis is performed on the recorded intensity profiles for quantifying the MMP directly from the collected fluorescence intensity data. In terms of speed, MMP can be obtained within 30 min

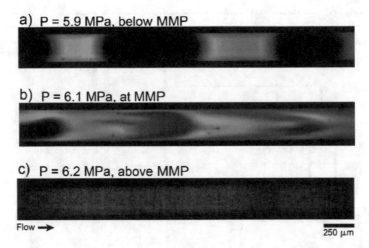

a) P = 5.9 MPa, below MMP

b) P = 6.1 MPa, at MMP

c) P = 6.2 MPa, above MMP

Flow ➡   ▬▬▬
                                                              250 μm

Fig. 5.2 Visualization of interfaces formed between CO_2 bubble and flowing bulk oil phase at different experimental pressure steps (**a** when pressure is below MMP, a clear interface is visible, **b** when pressure is equal to MMP, interface disappears at downstream location, [2] when pressure is above MMP, a uniform single phase is visible everywhere) modified figure reproduced by permission within the scope of Creative Commons 4.0 license: American Chemical Society (ACS), Fast fluorescence-based microfluidic method for measuring minimum miscibility pressure of CO_2 in crude oils, https://pubs.acs.org/doi/10.1021/ac5047856, Nguyen P, Mohaddes D, Riordon J et al. © 2015. Any further permissions related to the material excerpted should be directed to the ACS

using the microfluidic method developed by Nguyen et al. (2015), compared to days or week using other experimental methods.

There is no doubt that the fast fluorescence-based microfluidic MMP method developed by Nguyen et al. (2015) has the potential to inform and improve the state of the art of characterization and determination of CO_2-reservoir oil miscibility. However, its accuracy and reliability need to be further evaluated for the CO_2-reservoir oil systems that may exhibit condensing or mixed (vaporizing/condensing) drive mechanisms for attaining complete multicontact miscibility.

On the other hand, its capabilities can be further improved if information on the prevailing drive mechanism(s) can be directly inferred from the collected data. What if the observed fluctuations in fluorescence intensity can also be correlated in some manner with the vaporization of hydrocarbon components by injected CO_2 bubbles and the condensation of CO_2 into flowing oil phase? If such a strategy proved to be successful, then a fast fluorescence-based microfluidic MMP method, like the VIT method, may also provide direct and quantitative evidence on the prevailing drive mechanism(s) responsible for the development of miscibility in a given CO_2-reservoir system.

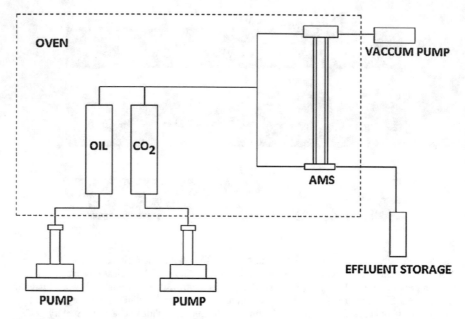

Fig. 5.3 Schematics of the experimental setup used in the acoustically monitored separator (AMS)-based sonic response method, reprinted by permission: Elsevier, Determination of minimum miscibility pressure for CO_2 and oil system using acoustically monitored separator. J CO_2 Utiliz 17:32–36, Czarnota R, Janiga D, Stopa J et al. © 2017

5.3.2 The Sonic Response Method

Another fast and easier to perform experimental method that has been recently reported in the published literature is the Sonic Response Method (SRM). The SRM method, which investigate the phase behavior changes between CO_2 and reservoir oil by using acoustically monitored separator (AMS), was developed by Czarnota et al. (2017a). The experimental setup for this AMS-based sonic response method is shown in Fig. 5.3.

The key feature of the experimental setup is a two-phase acoustically monitored separator (AMS), which includes two long (~51 cm) tubes with a diameter of 1.27 cm, mounted vertically in a stable base. The innovative design of AMS allows simultaneous separation of two mixing phases (i.e., CO_2 and reservoir oil) and measurement of the interface height. An acoustic transducer (i.e., a piezoelectric crystal attached to a titanium diaphragm) is used for generating acoustic waves that travel through the two mixing phases. The presence or absence of interface between the phases (i.e., CO_2-rich and oil-rich phases), as shown in Fig. 5.4, is detected by comparing the travel time of the reflected pulse from the interface and from a known point located in the oil-rich phase. As the system approaches the MMP, the formation of a single fluid phase of uniform density (i.e., disappearance of interphase or boundary between the two phases) makes it impossible to detect any reflected acoustic wave signal.

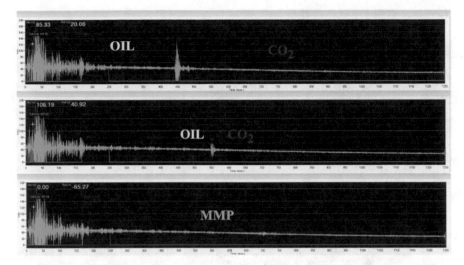

Fig. 5.4 Example of weakening acoustic signal during the determination of MMP for a CO_2-oil system at 23 °C by using acoustically monitored separator (AMS)-based sonic response method, reprinted by permission: Elsevier, Determination of minimum miscibility pressure for CO_2 and oil system using acoustically monitored separator. J CO_2 Utiliz 17:32–36, Czarnota R, Janiga D, Stopa J et al. © 2017

Czarnota et al. (2017b) also compared the MMP results obtained from the sonic response method with other experimental approaches including the slim-tube method and the rising bubble method as well. They were found in good agreement. According to the authors, the main advantage of the sonic response method is its ability to provide fast and reliable results without any encroachment into system.

5.3.3 The Rapid Pressure Increase Method

The main principle behind the rapid pressure increase (RPI) method as developed by Czarnota et al. (2017a) is the establishment of a relationship between pressure increase and volume reduction of binary system (i.e., CO_2 and reservoir oil) to detect the moment at which derivation of the established relationship is the lowest. The experimental setup for the RPI method is shown in Fig. 5.5.

During the measurement process, as depicted in Fig. 5.6, both the CO_2 and reservoir oil-filled piston-type accumulators are connected through an operation valve. Both fluids are allowed to come into contact through this operation valve. The pressure of both fluid accumulators is then increased by injecting the working fluid at a constant rate (e.g., 5 cc/min) using the available pump. The pumping of working fluid results in a gradual increase in the system's pressure due to decreased volume compressibility resulting from an increased solubility of phases into each other. A

(a)   **(b)**

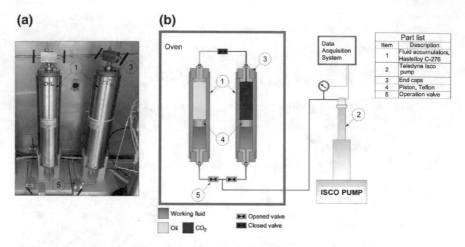

Fig. 5.5 Apparatus (**a**) and schematics diagram (**b**) of the experimental setup use in the rapid pressure increase method, reprinted by permission: Elsevier, Minimum miscibility pressure measurement for CO_2 and oil using rapid pressure increase method. J CO_2 Utiliz 21:156–161, Czarnota R, Janiga D, Stopa J et al. © 2017

corresponding change in volume with increasing pressure over time is recorded. In the beginning, due to sufficient solubility of both phases into each other, for a small increase in pressure, a significant reduction in volume is observed (Fig. 5.7, from point a to point c). Over time, as both phases become increasingly soluble (i.e., achieve miscibility), a sharp increase in pressure can be observed for a negligible volume reduction (Fig. 5.7, point c to point d). The slope of the tangent line drawn at the inflection point, represents the critical ratio of pressure change with respect to volume change at which a sharp increase in pressure is observed for a small volume change. The inflection point, which signifies the achievement of miscibility between the two phases due a loss of volume compressibility resulting from the formation of a single phase, is considered the MMP (Fig. 5.7, point c).

As reported by Czarnota et al. (2017a), the RPI-measured MMPs for certain CO_2-hydrocarbon systems are in good agreement with the reported MMPs obtained from other experimental approaches. According to Czarnota et al. (2017a), their method generates quick and plausible results during the experiment. However, its reliability in determining and characterizing the CO_2-reservoir oil miscibility in terms of the MMP for variety of CO_2-reservoir oil system still needs to be tested.

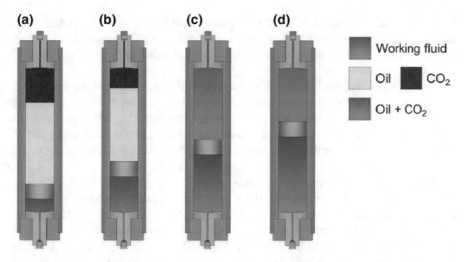

Fig. 5.6 Schematic depiction of measurement process (i.e., change in sample volume with increasing pressure) used in the rapid pressure increase method, reprinted by permission: Elsevier, Minimum miscibility pressure measurement for CO_2 and oil using rapid pressure increase method. J CO_2 Utiliz 21:156-161, Czarnota R, Janiga D, Stopa J et al. © 2017

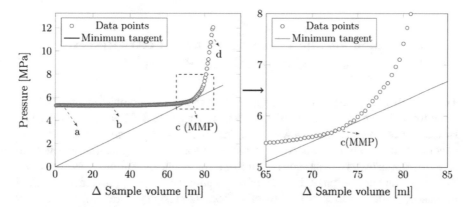

Fig. 5.7 Determination of the MMP from the experimental data (i.e., pressure versus change in sample volume relationship) collected by using the rapid pressure increase method, reprinted by permission: Elsevier, Minimum miscibility pressure measurement for CO_2 and oil using rapid pressure increase method. J CO_2 Utiliz 21:156–161, Czarnota R, Janiga D, Stopa J et al. © 2017

5.4   Recent Developments in Numerical Approaches

5.4.1   The Molecular Dynamic Simulation Method

Recently, Yang et al. (2016) presented the molecular dynamic simulation of miscible process that occurs between CO_2 and reservoir oil. The approach, which is based molecular simulation rather than physical simulation of phase behavior, relies on describing the movement and interactions of molecules for accurately depicting the mechanism(s) responsible for the development of miscibility between CO_2 and reservoir oil.

According to Yang et al. (2016), the molecular dynamic simulation method could simulate any real or virtual system and could give direct forecast of many parameters of the system. For example, they used the method to predict the MMPs of selected CO_2-hydrocarbons mixtures, which, from the molecular dynamic simulation point of view, is considered to be critically influenced by the structure of aromatic, naphthenic, and paraffin hydrocarbons with different carbon number.

Based on the MMP prediction results obtained for simple mixtures of CO_2 and hydrocarbons (i.e., different hydrocarbons [paraffin, naphthenic, and aromatic] and different crude oil molecular chains [C_6, C_{10}, C_{20}]), Yang et al. (2016) postulate that the application of molecular dynamics simulation technology can provide another way to study the phase behavior in terms of molecular motion and force field, which can be subsequently used for studying the mechanism of phase behavior responsible for the development of miscibility at MMP. The results are encouraging; however, the capabilities of molecular dynamic simulation approach in predicting MMPs for CO_2-reservoir oil systems still need to be explored.

5.5 Determination and Characterization of CO_2-Reservoir Oil Miscibility in Unconventional Reservoirs

The oil recoveries in unconventional tight and/or shale oil reservoirs (e.g., the Bakken shale formation in North America) is still as low as 5–10% (Alfarage et al. 2018). Based on a comprehensive review of published literature Alfarage et al. (2018) have concluded that the most feasible EOR methods for recovering additional oil from unconventional reservoirs include miscible gas injection, low-salinity waterflooding, and surfactant application, with CO_2-EOR topping the list of applied methods tried in shale oil reservoirs to date. In view of growing interest in researching the viability of CO_2-EOR in unconventional reservoirs, a better understanding of the issue of determination and characterization of CO_2-reservoir oil miscibility in such reservoirs needs to be developed.

Some of the notable recent advancements that have been made in this direction include the use of the RBA method and the development of a newer version of the VIT method to experimentally determine the MMP for CO_2-tight reservoir oil systems. Apart from such experimental efforts, the development of new analytical and numerical methods (e.g., integrated MMP correlation, PC-SAFT EOS model, and diminishing interface method [DIM]) have enhanced our capability to accurately characterize and determine CO_2-reservoir oil miscibility in terms of the MMP for the design and implementation of CO_2 injection-based miscible EOR and/or EOR and CO_2 storage strategies in unconventional reservoirs.

5.5.1   The RBA Method

Recently, Li and Luo (2017) have reported on RBA method-measured MMPs for five candidate injection gases—pure CO_2, CO_2-enriched flue gas, natural gas, nitrogen, and CO_2-enriched natural gas—and Bakken live oil. The measured MMP data reveal that the pure CO_2-Bakken live oil system has much lower MMP than other gases tested. The measured MMP for this system was also found to be lower than the virgin pressures of most Bakken reservoirs located in the Canadian Province of Saskatchewan.

Similarly, Adekunle and Hoffman (2016) chose the RBA method when they evaluated the feasibility of miscible gas injection in the Bakken shale formation located in the US portion of Williston Basin (i.e., North Dakota and Montana). The measured MMP for both dead and recombined live Bakken oil samples while using pure gases of CO_2, N_2, CH_4, and two enriched hydrocarbon gas streams (CH_4 and C_3H_8 mixed in 80/20 and 60/40 mol ratios). Again, pure CO_2 achieved miscibility at the lowest pressure compared to all three hydrocarbon gas streams, whereas miscibility was not achieved with N_2. The MMPs for recombined live Bakken oil were found to be higher than the MMPs for dead oil expected when CH_4 was used as the injection gas phase.

5.5.2   The Capillary-Rise-Based VIT Method

Hawthorne et al. (2016) reported the use of a simplified version of the VIT method for determining the MMP for CO_2-Bakken tight live reservoir oil system. The method, referred to here as the capillary-rise-based VIT method, simply measures the height of an observed rise of CO_2-reservoir oil interface in three capillary tubes of varied diameters that are housed in a high-pressure view cell, for inferring the extent of miscibility between the two equilibrated (i.e., oil-rich and CO_2-rich) phases at a given experimental pressure and reservoir temperature. As discussed earlier, the presence of an interface between the two equilibrated phases signifies the immiscible nature of the system at experimental pressure steps below MMP. As the system approaches the attainment of miscibility at experimental pressure steps close to MMP, the interface between the two equilibrated phases vanishes. The same is manifested by the diminished height of an observed rise of CO_2-reservoir oil interfaces in the capillary tubes.

Hawthorne et al. (2016) extrapolated the height of capillary-rise versus pressure plots to zero height for obtaining the MMP for the CO_2-Bakken tight live reservoir oil system. According to them, the mean MMP deduced from five capillary-rise-based measurements was in good agreement with the experimentally-measured MMP value (the slim-tube method).

According to Hawthorne et al. (2016), the main advantage of their method is that MMP can be determined without the need of the additional instrumentation and labor required to perform actual IFT measurements. However, it may not be necessarily the case. Let's take a closer look at the IFT and capillary rise height relationship using the well-known governing equation of capillary rise height given below. The same has been used by researchers (e.g., Ayirala 2005; Ayirala and Rao 2006; Sequeira 2006; Sequeira et al. 2008; Saini and Rao 2010) for experimentally measuring the CO_2-reservoir oil IFTs at various experimental steps and reservoir temperature.

$$\gamma = \frac{rh(\rho_l - \rho_v)g}{2cos\theta} \tag{5.1}$$

where, γ is the IFT (dynes/cm) between oil and gas phase and ρ_l and ρ_v are the densities (g/cm^3) of liquid and gas phase, respectively. r is the internal radius of capillary tube (cm), h is height of capillary rise (cm), $cos\ \theta$ is the equilibrium contact angle in degrees, and g (cm/s^2) is acceleration due to gravity.

As evident from Eq. 5.1, γ is proportional to both h and (ρ_l -ρ_v). It is well known (e.g., Figures 3.13 through 3.16) that, as the system approaches towards the attainment of miscibility, density differences between the equilibrated phases disappear rapidly in a non-linear fashion. In other words, the use of a linear relationship between γ and h and while neglecting the non-linear effect of density difference (ρ_l -ρ_v) on γ may lead to an erroneous MMP value. For example, Saini (2016a) compared the MMPs for one of the CO_2-reserovir oil systems (i.e., 89 + 11) investigated by Sequeira (2006); Sequeira et al. 2008 (Table 3.2) from the extrapolation of the experimentally-measured IFT and pressure (i.e., γ versus P) curves (Fig. 5.8a) and the experimentally-measured IFT versus 1/pressure (i.e., γ and 1/P) curve (Fig. 5.8b) with the experimentally-observed relationship between the height of capillary rise versus pressure (i.e., h and P) (Fig. 5.8c) and the height of capillary rise versus 1/pressure (i.e., h and 1/P) curves to the conditions of zero IFT. As can be seen from the comparison of the MMP values obtained from Figs. 5.8a through 5.8d, the difference between the MMP value inferred from γ versus 1/P curve (Fig. 5.8b) and the MMP value inferred from h versus P relationship (Fig. 5.8c) was found to be 12%.

On the other hand, a comparison of the MMP values obtained from two distinct trends observed in IFT versus pressure data (Fig. 5.8a) with the MMP value obtained from h versus $1/P$ (Fig. 5.8d), resulted in a difference ranging from 200 to 349%. As can be seen in Fig. 5.8d, at an experimental pressure step close to the MMP, the height of capillary rise suddenly dropped to a very low value, which provides a visual evidence that system was about to achieve complete miscibility. However, the extrapolation of the linear trend observed in the height of capillary rise versus pressure relationship (Fig. 5.8c) to a condition to zero IFT may still lead to erroneous result because at experimental pressure steps near to the MMP, such linear trend is going to vanish as the system approaches the miscibility. Hence, the capillary-rise-based VIT method presented by Hawthorne et al. (2016), which can still be used for making a quick and qualitative initial estimate of the MMP, may not provide an accurate quantitative estimate of the MMP. For quantitative estimation, either densities of the equilibrium oil-rich and CO_2-rich phases are measured, or the absence of interface manifested by zero height of the capillary rise is confirmed experimentally.

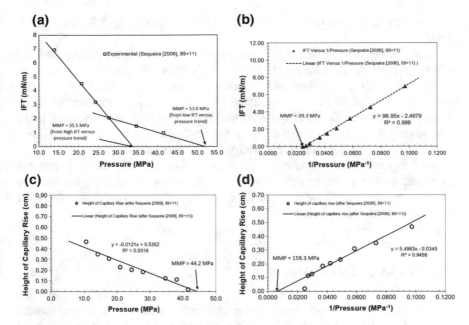

Fig. 5.8 Comparison of the MMP values derived from the IFT versus pressure (**a**) and the IFT versus 1/pressure curves with the MMP values derived from the height of capillary rise versus pressure (**c**) the height of capillary rise versus 1/pressure curves at 114.4 °C (238 °F) for a CO_2-live reservoir oil system (89 + 11, Table 3.2) investigated by Sequeira (2006); Sequeira et al. (2008)

5.5.3 Integrated MMP Correlation

As highlighted by Zhang et al. (2016), in a confined system (i.e., nanoscale pores of tight oil reservoirs), the phase equilibrium and other fluid properties of both the reservoir oil and injected CO_2 may deviate from those at their bulk phase. It may also result in a decreased MMP with confinement. Hence, it is necessary that existing correlations are revised to make initial approximate predictions of the MMP for CO_2-tight oil reservoir oil systems.

To address this issue, Zhang et al. (2016) first investigated the applicability of existing MMP correlations in predicting MMP for CO_2-tight reservoir oil systems; however, none of them was found suitable to include confined effect. Afterwards, they proposed a new solubility-based integrated MMP correlation that is capable of taking the confinement effect into consideration. Zhang et al. (2016) has provided a detailed discussion of the methodology that they had used in deriving this new integrated MMP correlation. In nutshell, their methodology relies on the determination of CO_2 density after taking the confinement effect into consideration via the incorporation of a critical property shift. The calculated value of CO_2 density with confinement can then be used for determining the revised MMP from an available MMP versus

CO_2 density relationship. One such relationship can be seen in Fig. 3 by Zhang et al. (2016).

For validating their integrating MMP correlation, Zhang et al. (2016) used the following approach. First, they used five live tight oil samples including Cardium, Bakken, Monterey, Eagle Ford, and Niobrara reservoirs, for validating commonly used MMP correlations (Table 1 of Zhang et al. 2016) using a commercial phase behavior package (CMG WINPROP). Subsequently, they predicted the MMP using their integrated correlation at two different pore radii (50 and 10 nm) and validated the results by comparing them with CMG WINPROP predicted MMPs.

The integrated MMP correlation presented by Zhang et al. (2016) is one the key recent advancements made towards enhancing our prediction capabilities for screening the unconventional reservoirs for their suitability in implementing a CO_2 flooding project in near future.

5.5.4   PC-SAFT EOS Model

The use of both the basic Parachor and the mechanistic Parachor models coupled with EOS models for predicting the MMP for CO_2-reservoir oil systems has been reported in published studies. For example, Nobakht et al. (2008) modeled the measured equilibrium IFT behaviors of a CO_2-Weyburn crude oil system using the Parachor model and the linear gradient theory (LGT) model. They used the EOS generated vapor-liquid equilibrium (VLE) data for predicting the IFT behaviors using the Parachor model and the LGT model. Ashrafizadeh and Ghasrodashti (2011) presented a comparative study of five representative EOSs for predicting MMP using the mechanistic Parachor model together with the criterion of zero IFT at the miscibility conditions. According to them, all of the studied EOSs can predict MMP using the mechanistic Parachor model within an acceptable range of accuracy however the model might be found non-suitable for scenarios which incorporate considerable amounts of polar components. In such cases, they suggested the use of more complicated EOSs such as electrolyte models.

Tecklu et al. (2014) reported the development of a computational approach for determining the MMP which mimics the VIT method. In their approach, they used the Parachor model (Zuo and Stenby 1998) for modeling the VIT trend of equilibrium gas-oil with increasing pressure. They mentioned that other Parachor models such as the mechanistic Parachor model (Ayirala and Rao 2006) can also be used for obtaining the VIT trend. It is worth mentioning here that such studies essentially rely on an approach which is very similar to the simulated VIT method. As discussed earlier, in the simulated VIT method, a well-tuned EOS model is used for obtaining certain input data required for determining the pressure dependence of the IFT behavior of a given CO_2-reservoir oil system using either the basic Parachor or the mechanistic Parachor model. The calculated IFT versus pressure curve is then extrapolated to the condition of zero IFT (i.e., MMP).

Recently, Wang et al. (2016) extended the capabilities of the basic Parachor model to account for the confinement effect of nanopore walls on the equilibrium IFT. For generating the phase equilibrium data (i.e., equilibrium liquid and vapor phase compositions and densities) between reservoir oil and injected gas, which are necessary input parameters for deriving the pressure dependence of IFT behavior in nanopores, they used PC-SAFT EOS model. The modified Parachor model coupled with PC-SAFT model was then used for predicting the MMP between Bakken reservoir oil and CO_2 using the simulated VIT method.

When compared with the bulk phase, the calculated MMP for CO_2-Bakken reservoir oil system was found to be suppressed by approximately 2.6% in 10 mm pore and 23.5% in 3 nm pores. Also, according to Wang et al. (2016), the MMP for CO_2-Bakken reservoir oil system is more or less independent of pore width when pore width is greater than 10 nm.

These findings are quite encouraging from the point of view of CO_2-EOR as miscibility condition in nanopores can be achieved at lower injection pressures. Obviously, the achievement of miscibility will assist in reducing the flow resistance in such pores. However, in fractured regions (whether natural or hydraulically induced) of the reservoir, miscibility between injected CO_2 and reservoir oil is still expected to be achieved at higher pressures due to high MMPs in the bulk phase. Irrespective of observed MMP variations in bulk phase and in nanopores, application of miscible CO_2-EOR can still turn out to be a viable option for recovering additional oil from unconventional reservoirs like the Bakken formation.

5.5.5 The Diminishing Interface Method

The diminishing interface method (DIM) presented by Zhang et al. (2017) uses both the simulated VIT method, which is based on the Parachor model coupled with a modified PREOS model, and a formula for determining the interfacial thickness between two mutually soluble phases for determining the MMPs of CO_2-tight reservoir oil system in bulk phase and nanopores. For determining the MMP from the diminishing interface method, the derivative of the calculated interface thickness with respect to pressure is extrapolated to zero. Physically, attainment of zero derivative means that, at pressures $\geq$MMP, the interfacial thickness between the CO_2-rich and the oil-rich phases becomes constant and it does not change with pressure.

According to Zhang et al. (2017), the PREOS model, which is modified to calculate the VLE in nanopores by considering the effects of capillary pressure and shifts of critical temperature and pressure, when coupled with the Parachor model can accurately predict the pressure dependence of IFT behaviors of CO_2-reservoir oil systems in bulk phase as well as in nanopores. Similarly, the diminishing interface method can also accurately predict the MMP.

For deducing this conclusion, Zhang et al. (2017) compared the simulated VIT method-derived MMPs, the diminishing interface method-derived MMPs, and the VIT method-measured MMPs for three CO_2-reservoir oil systems—namely the CO_2-

Pembina dead light reservoir oil, the CO_2-Pembina live light reservoir oil, and the impure CO_2-Pembina dead light reservoir oil systems. Both the VIT and the simulated VIT MMPs were also compared with the MMPs measured using other experimental methods including the coreflood tests (CO_2-Pembina dead light reservoir oil system), the slim-tube method (CO_2-Pembina live light reservoir oil system), and the RBA method (impure CO_2-Pembina dead light reservoir oil system). The MMPs obtained from different approaches reasonably agreed. It is noted that, in this text, both the terms "light reservoir oil" and "tight reservoir oil" have been used interchangeably to describe high API gravity (i.e., >37° or more) crude which is generally encountered in unconventional reservoirs like the Bakken formation.

One major implication of the MMP results obtained by using the diminishing interface method is that they support the notion of the presence of a stable interfacial thickness between the oil and CO_2 phases at MMP rather than the complete disappearance of interface, which is generally honored in the VIT method or the simulated VIT method by extrapolating the measured or the simulated IFT data to the condition of zero IFT for determining the MMP. However, it is interesting to note that, for all three CO_2-reservoir oil systems investigated by Zhang et al. (2017), the IFT versus pressure curves derived using the simulated VIT method (i.e., Parachor model coupled with modified PREOS model) and experimentally-measured IFT versus pressure data (i.e., the VIT method), as shown in Fig. 5 by Zhang et al. (2017), appear to be in good agreement only in a high IFT region (i.e., IFT > 3 mN/m). In a low IFT region (i.e., <3 mN/m), a significant deviation between the simulated IFTs and the experimental IFTs can easily be observed.

Zhang et al. (2017) measured the equilibrium densities for an oil-rich phase experimentally; however, the equilibrium densities for CO_2-rich phase were calculated by using a commercial phase behavior package (CMG WINPROP). Hence, it is possible that the lack of use of measured equilibrium densities for CO_2-rich phase has caused this apparent disagreement between the simulated IFTs and the experimental IFTs in low IFT regions. As demonstrated by Saini (2016a), if experimentally measured equilibrium phase densities are used, in low IFT region, a better agreement between the simulated IFTs and the experimental IFTs can be obtained.

In view of the above-mentioned discussion, the notion of the presence of a stable interfacial thickness between the oil and CO_2 phases at MMP rather than the complete disappearance of interface as inferred from the diminishing interface method needs to be explored further by using experimental techniques including the one described by Danesh (1998). He described a way to measure very low interfacial tension values by measuring the interface curvature which appears as a band with a finite thickness between phases due to the light scattering in an optical cell.

References

Adel IA, Tovar FD, Schechter DS (2016) Fast-slim tube: a reliable and rapid technique for the laboratory determination of MMP in CO_2-light crude oil systems. Soc Pet Eng. https://doi.org/10.2118/179673-MS

Adekunle O, Hoffman BT (2016) Experimental and analytical methods to determine minimum miscibility pressure (MMP) for Bakken formation crude oil. J Pet Sci Eng 146:170–182

Al Riyami M, Rao DN (2015) Estimation of near-miscibility conditions based on gas-oil interfacial tension calculations. Soc Pet Eng. https://doi.org/10.2118/174646-MS

Alfarge D, Wei M, Bai B (2018) Data analysis for CO_2-EOR in shale-oil reservoirs based on a laboratory database. J Pet Sci Eng 162:697–711

Ashrafizadeh SN, Ghasrodashti AA (2011) An investigation on the applicability of Parachor model for the prediction of MMP using five equations of state. Chem Eng Res and Des 89(6):690–696

Ayirala SC (2005) Measurement and modeling of fluid-fluid miscibility in multicomponent hydrocarbon systems. Thesis, Louisiana State University

Ayirala SC, Rao DN (2006) Comparative evaluation of a new MMP determination technique. Soc Pet Eng. https://doi.org/10.2118/99606-MS

Czarnota R, Janiga D, Stopa J et al (2017a) Determination of minimum miscibility pressure for CO_2 and oil system using acoustically monitored separator. J CO_2 Utiliz 17:32–36

Czarnota R, Janiga D, Stopa J et al (2017b) Minimum miscibility pressure measurement for CO_2 and oil using rapid pressure increase method. J CO_2 Utiliz 21:156–161

Danesh A (1998) PVT and phase behavior of petroleum reservoir fluids. Elsevier Science B.V, Netherlands

Hawthorne SB, Miller DJ, Jin L (2016) Rapid and simple capillary-rise/vanishing interfacial tension method to determine crude oil minimum miscibility pressure: pure and mixed CO_2, methane, and ethane. Energy Fuels 30(8):6365–6372

Li S, Luo P (2017) Experimental and simulation determination of minimum miscibility pressure for a Bakken tight oil and different injection gases. Petroleum 3(1):79–86

Mogensen K (2016) A novel protocol for estimation of minimum miscibility pressure from slimtube experiments. J Pet Sci Eng 146:545–551

Nguyen P, Mohaddes D, Riordon J et al (2015) Fast fluorescence-based microfluidic method for measuring minimum miscibility pressure of CO_2 in crude oils. Anal Chem 87(6):3160–3164

Nobakht M, Moghadam S, Gu Y (2008) Determination of CO_2 minimum miscibility pressure from measured and predicted equilibrium interfacial tensions. Ind Eng Chem Res 47:8918–8925

Saini D (2016a) An investigation of the robustness of physical and numerical vanishing interfacial tension experimentation in determining CO_2 +crude oil minimum miscibility pressure. J Petro Eng 2016:13, https://doi.org/10.1155/2016/8150752

Saini D, Rao DN (2010) Experimental determination of minimum miscibility pressure (MMP) by gas/oil IFT measurements for a gas injection EOR project. Soc Pet Eng. https://doi.org/10.2118/132389-MS

Sequeira DS (2006) Compositional effects on gas-oil interfacial tension and miscibility at reservoir conditions. Thesis, Louisiana State University

Sequeira DS, Ayirala SC, Rao DN (2008) Reservoir condition measurements of compositional effects on gas-oil interfacial tension and miscibility. Soc Pet Eng. https://doi.org/10.2118/113333-MS

Teklu TW, Alharthy N, Kazemi H (2014) Phase behavior and minimum miscibility pressure in nanopores. Soc Pet Eng. https://doi.org/10.2118/168865-PA

Wang S, Ma M, Chen S (2016) Application of PC-SAFT equation of state for CO_2 minimum miscibility pressure prediction in nanopores. Soc Pet Eng. https://doi.org/10.2118/179535-MS

Wang H, Lun Z, Lv C et al (2017) Measurement and visualization of tight rock exposed to CO_2 using NMR Relaxometry and MRI. Sci Rep 7. http://dx.doi.org/10.1038/srep44354

Yang S, Lian L, Yang Y et al (2016) Molecular dynamics simulation of miscible process in CO_2 and crude oil system. Soc Pet Eng. https://doi.org/10.2118/182907-MS

Zhang K, Jia N, Zeng F et al (2017) A new diminishing interface method for determining the minimum miscibility pressures of light oil–CO_2 systems in bulk phase and nanopores. Energy Fuels 31(11):12021–12034

Zhang K, Seetahal S, Alexander D et al (2016) Correlation for CO_2 minimum miscibility pressure in tight oil reservoirs. Soc Pet Eng. https://doi.org/10.2118/180857-MS

Zuo Y-X, Stenby EH (1998) Prediction of interfacial tensions of reservoir crude oil and gas condensate systems. Soc Pet Eng. https://doi.org/10.2118/38434-PA

Printed in the United States
By Bookmasters